Michael Chiappini -1997-

W9-CQX-079

PLANTS FROM
TEST TUBES

PLANTS FROM TEST TUBES

An Introduction to Micropropagation

Third Edition

LYDIANE KYTE AND JOHN KLEYN

TIMBER PRESS
Portland, Oregon

This edition is dedicated to the memory of Dr. Hudson Hartmann (1915–1994).
His untiring efforts to promote good horticulture through his labors as professor and author
and as mentor for the International Plant Propagators' Society
have made a lasting impression on the state of the art.

Copyright © 1996 by Timber Press, Inc.
All rights reserved.

ISBN 0-88192-361-3

Printed in Hong Kong

Timber Press, Inc.
The Haseltine Building
133 S.W. Second Avenue, Suite 450
Portland, Oregon 97204, U.S.A.

Library of Congress Cataloging-in-Publication Data

Kyte, Lydiane.
Plants from test tubes : an introduction to micropropagation /
Lydiane Kyte and John Kleyn. —3rd ed.
p.   cm.
Includes bibliographical references (p.   ) and index.
ISBN 0-88192-361-3
1. Plant micropropagation—Laboratory manuals.
2. Plant tissue culture—Laboratory manuals.
I. Kleyn, John G. II. Title
SB123.6.K98   1996
631.5'3—dc20   96-33972
CIP

Contents

Foreword by Michael E. Kane   7
Preface   9

SECTION I   THE BASICS OF TISSUE CULTURE
  Chapter 1   Overview   13
  Chapter 2   The Botanical Basis for Tissue Culture   21
  Chapter 3   History   27
  Chapter 4   The Laboratory—Facilities and Supplies   35
  Chapter 5   Media Ingredients   60
  Chapter 6   Media Preparation   76
  Chapter 7   Explants and Their Preparation   84
  Chapter 8   The Transfer Room   90
  Chapter 9   The Growing Room   101
  Chapter 10   Hardening Off   108
  Chapter 11   Dealing with Biological Contaminants   113
  Chapter 12   Business   137
  Chapter 13   Biotechnology   146

SECTION II   CULTURE GUIDE TO SELECTED PLANTS
  Ferns   169
  Conifers   170
  Monocotyledons   171
  Dicotyledons   183

Formula Comparison Chart   206
Appendices
  A. Metric Conversions   208
  B. The Microscope—Introduction, Use, and Care   210
  C. Professional Organizations   214
  D. Suppliers   215
Glossary   219
Bibliography   227
Index   239

Foreword

More than 30 years have elapsed since George Morel first proposed the application of plant tissue culture for the commercial clonal propagation of virus-free orchids. Since that time, *in vitro* propagation, or micropropagation as it is commonly known, has evolved into a competitive, worldwide industry that produces 250 million plants annually. In the United States alone there are approximately 75 large commercial micropropagation laboratories. Production has long centered on high-value ornamental crops, such as tropical foliage plants and orchids. More recently, commercial micropropagation of woody landscape plants, fruits, vegetables, and specialty crops has become increasingly common as laboratories strive to fill new markets. Undoubtedly, micropropagation has proven to be a commercially viable method for plant production and an important tool in the application of plant biotechnology in agriculture. Scientific advancements are continually being made in areas such as synthetic seed technology and automated micropropagation systems development. However, the information being generated is often available only in very expensive and specialized journals, symposium proceedings, and advanced text books. Unfortunately, the number of affordable texts that provide a practical but thorough overview of the micropropagation process have been few. For those who ask, What is micropropagation and how can I do it?, the choices have been limited.

Since it was first published in 1983, Ann Kyte's *Plants from Test Tubes* has served as an affordable and useful primer for the plant propagator, student, and curious gardener who wanted to learn the practical "nuts and bolts" of commercial micropropagation. No other book has filled this niche as well as *Plants from Test Tubes*. Ann Kyte's real-world experience as manager and owner of a micropropagation laboratory has enabled her to write a user-friendly book covering all aspects of commercial production. This expanded third edition will ensure that future "budding" students of micropropagation will have a solid introduction to this fascinating and rapidly evolving technology!

Michael E. Kane
University of Florida

Preface to Third Edition

The first edition of *Plants from Test Tubes* came about as a result of the realization by Timber Press that there was a need for a tissue culture primer that could serve the general public and commercial horticulture, areas where tissue culture was not yet widely practiced. The response was overwhelming. The book found its way beyond the United States and Canada, appearing in Egypt, in Brazil, in Australia, in India, and in Europe, where it was even translated into Danish.

The basics of tissue culture have changed very little in the past 20 years. A reliance on sterile technique and on variations of Murashige and Skoog's medium formula continue to be focal points of successful micropropagation. But even in the academic world, where sterile technique is commonplace and articles in technical journals are readily available, the book has proven useful as an introductory text.

This third edition was launched due to an outcry of demand when the second edition went out of print. (The demand echoed in a tissue culture network.) It seemed appropriate to widen the scope of the book and to touch on the explosion in the related field of biotechnology, a field that depends on the use of tissue culture in such areas as the multiplication of transformed plants. Cell culture is also introduced in this edition, as further research suggests the numerous applications of that procedure in industry.

Without the encouragement and contribution of Dr. John Kleyn, my collaborator, this edition would not even have been attempted. As a microbiologist, professor, commercial tissue culturist, and consultant, he brings a new dimension to the book. Specifically, he suggests to the reader methods of preventing and identifying microbial contaminants, the bane of all tissue culture.

For their encouragement and helpful suggestions I would like to thank the many devotees of previous editions and my associates in the International Plant Propagators' Society. John Kleyn joins me in appreciation of artist Sandy Godsey, and gratitude for the patience of our respective spouses, Bob Kyte and Jan Kleyn.

Lydiane Kyte
Centralia, Washington

SECTION I

The Basics of Tissue Culture

Chapter 1: Overview

Plant tissue culture is a field of many facets. Its applications vary from the curious gardener multiplying plants in a modest home kitchen, to the renowned scientist working in an elaborate laboratory. It reaches from the orchid hobbyist who has learned to multiply a few personal favorites, to a million-dollar industry producing houseplants, ornamentals, and secondary products. Tissue culture is the ever-ready tool for specialists who hybridize plants by either sexual or asexual means. It is a clean and rapid way for genetic engineers to grow material for identifying and manipulating genes or to transfer individual characteristics from one plant to another. It plays a role in a wide array of fields, such as botany, chemistry, physics, genetic engineering, molecular biology, hybrid development, pesticide testing, and food science.

The number of nurseries engaged in plant propagation by tissue culture rapidly escalated during the 1970s and 1980s. The basic scientific principles of plant tissue culture were established—one might say that commercial plant tissue culture had come of age. This exciting technique was a result of active research and was employed by progressive growers, students, hobbyists, and gardeners, while still serving as a valuable tool for the plant scientist.

Hundreds of commercial laboratories are currently involved in micropropagation (as tissue culture may be more aptly known) with some laboratories producing over 20 million plants a year. Previously there had been doubts about the feasibility of using tissue culture to propagate woody plants (trees and shrubs), but it is estimated that by the mid-1980s "woodies" accounted for about 16% of total worldwide tissue culture production (George and Sherrington 1984).

Tissue culture was first used on a large scale by the orchid industry in the 1950s. Some fortuitous early discoveries opened the door to tissue culture for quality orchids, where previously growers had struggled with unpredictable seed or difficult-to-propagate, virus-infected stock. Later, it became clear that any plant would respond to tissue culture as long as the right formula and the right processes were developed for its culture.

With the widespread acceptance of the technique in the nursery business, it is surprising how many people are still not aware of plant tissue culture. Some people ask, when they hear or read about tissue culture, What is it? How is it done? Who is doing it and why? You may ask, Can I do this myself? or, Should I even try? What are the costs of building a laboratory? What are the potential financial returns? Some want to know who invented it or where it came from. This chapter and the ones that follow are designed to help answer some of these questions.

Although the science of botany has become increasingly complex with the explosion in the field of biotechnology, the procedures of tissue culture are not complicated (Figure 1-1). A piece of a plant, which can be anything from a piece of stem, root, leaf, or bud to a single cell, is placed in that tiniest of greenhouses, a test tube. In an environment free from microorganisms and in the presence of a balanced diet of chemicals, that bit of plant, called an *explant*, can produce plantlets that, in turn, will multiply indefinitely, if given proper care. The *medium* (plural, *media*) is the substrate for plant growth, and in the context of plant tissue culture it refers to the mixture of certain chemical compounds to form a nutrient-rich gel or liquid for growing cultures, whether cells, organs, or plantlets.

The process of tissue culturing plants from the explant stage to the final stage of transferring a mature plant to field or greenhouse conditions involves 4 basic stages. The 4 stages of culture growth are: Stage I, explant establishment or initiation; Stage II, multiplication; Stage III, root-

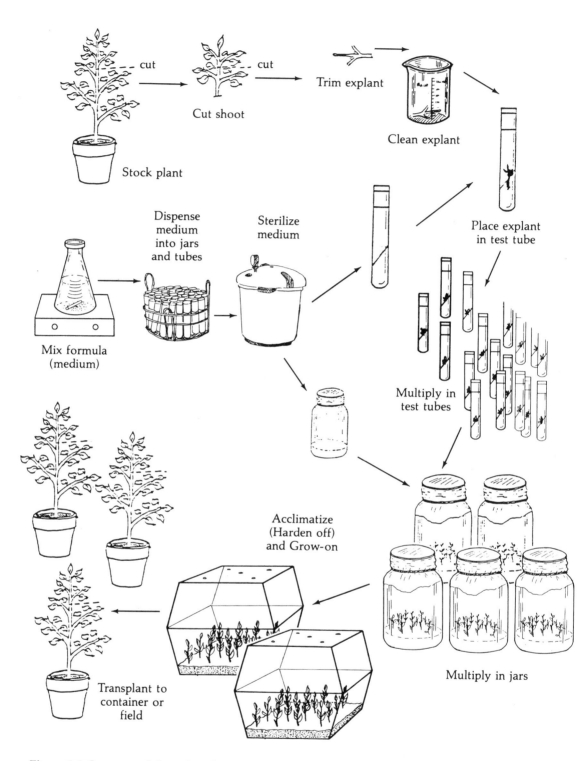

Figure 1-1. Sequence of shoot tip micropropagation.

ing; and Stage IV, acclimatization or hardening off. These stages can overlap in certain cases, and the requirements of each stage vary widely from plant to plant.

When new material is started in culture, grown *in vitro* (literally, "in glass"), it develops very small juvenile shoots, which are reminiscent of seedlings. A plantlet continues to produce and maintain small stems and leaves throughout its duration in culture. This is fortunate because most mature material would be too unwieldy for micropropagation to succeed in a test tube (Figure 1-2). After multiplication in culture and when transferred to soil outside the laboratory, the plantlets will produce leaves of normal size and assume the mature features of the plants from which they originated.

When plants are multiplied vegetatively—as distinguished from those grown from seeds—whether by tissue culture or by cuttings, all the offspring from a single plant can be classified as a *clone*. This means that the genetic make-up of each offspring is identical to that of all the other offspring and to that of the single parent. On the other hand, plants propagated by seed, resulting from sexual reproduction, are not clones because each seed (and the resultant plant) has a unique genetic make-up—a mixture from 2 parents, different from either parent and different from one seed to another. The term "cloning," with respect to tissue culture, refers to the process of propagating in culture large numbers of selected plants with the same *genotype* (the same genes or hereditary factors) as their respective parent plant.

The term "tissue culture" is actually a misnomer borrowed from the field of animal tissue culture. It is a misnomer because plant micropropagation is concerned with the whole plantlet and not just isolated tissues, though the explant may be a particular tissue. The terms "plantlet culture" or "micropropagation," therefore, are more accurate. But whether we call it cloning, tissue culture, micropropagation, or growing *in vitro*, the process remains the same: it is a vegetative method for multiplying plants. This method of propagation is commonly accepted by the nursery trade and it has had a significant impact on commercial horticulture.

Embryo culture, cell culture, and callus culture are usually designated as such, but they may also fall under the broad term of tissue culture. *Embryo culture* can mean the "rescue" of an embryo from a seed and fostering its plantlet development and multiplication in a culture, or it can refer to *somatic embryogenesis*, whereby embryos are induced to form from *somatic* (vegetative, or asexual) cells. *Cell culture* is the cultivation of cells on a solid gel medium or in a liq-

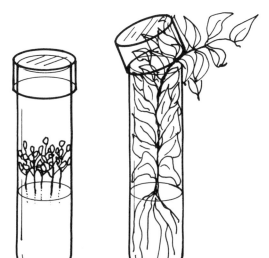

Figure 1-2. If tissue cultured plants did not revert to a juvenile state, most material would never fit into a test tube or jar and they would be too unwieldy to micropropagate.

uid medium, the latter commonly known as cell suspension culture. *Callus culture* is the multiplication of *callus* (a mass of disorganized, mostly undifferentiated or undeveloped cells), usually on a solid medium.

Some problems in plant propagation can originate with the stock plants from which seeds or cuttings are taken. Oftentimes, however, diseases that are normally transmitted from parent to offspring can be eliminated through tissue culture procedures. External contaminants, such as bacteria, fungi, spores, and insects, are removed when an explant is cleaned, and some internal contaminants, such as viruses, can be eliminated by using the apical meristem as the explant for the tissue culture. The *apical meristem* is the new, undifferentiated tissue at the microscopic tip of a shoot. It is often virus free even in diseased plants because these meristematic cells are not yet joined to the plant's vascular system, and perhaps they grow faster than the viruses. Thus, if the few virus-free cells that make up the microscopic dome of apical meristem are removed from the plant and placed in a culture, they can grow and produce healthy, disease-free plants. This technique is known as *meristem culture*, a term sometimes wrongly used to broadly denote micropropagation or tissue culture, but which should be limited to indicate cultures started from an apical meristem.

A valid concern does arise with respect to the genetic stability of tissue cultured plants. Thousands of plants could be cultured at one time, only to reveal some defect when they are planted and growing in the field—a defect resulting from a chemical imbalance or a mutation that was multiplied in culture. As a practical matter, however, cultures have been known to remain healthy and normal for decades, showing no aberrations. Some species are more prone to mutation than others. Cell and callus cultures are more likely to be genetically unstable than those grown as plantlets. With limited experience in a particular species, a tissue culturist should start with numerous explants, transfer them frequently, limit the number of subcultures, and start with new material annually or as often as experience dictates.

Tissue culture can serve a number of purposes, and growers have started their own commercial production laboratories for a variety of reasons. Growing plants from seeds or cuttings can be unacceptable or impractical due to some of the following (and other) factors:

- seed-grown products lack uniformity
- seed-grown products are not true to type
- seeds take too long to grow to mature plants
- seeds are difficult to handle
- seeds are not available
- cuttings grow too slowly
- cuttings have poor survival rate
- cuttings require too much care
- cuttings are too vulnerable to disease
- there is a shortage of stock plants from which to take cuttings because there is:
    only one hybrid
    only one virus-free plant
    only one desirable mutant
- there is insufficient room for stock plants
- it is not cost effective to maintain the stock plants

Tissue culture is often the only practical way to produce the large numbers required.

If an ample supply of seeds is available, or if plants from cuttings are acceptable and cutting stock is available and not a problem to maintain, then tissue culture may not be the most practical option because it can be an expensive, labor-intensive process, especially if less than a thousand plants are needed. If plants from seeds are acceptable but there are not enough seeds, then seeds or excised embryos can be used as starting material for tissue culture.

Every year excessive amounts of growers' time, labor, and space are spent on unproductive seeds, cuttings, and grafts. Significant numbers of young plants are lost to pests, diseases, or other environmental factors. Tissue cultured plantlets are less subject to such attacks and disasters because in the sterile environment of the laboratory they are not exposed to the pathogens or extreme conditions that afflict many plants grown in the field or greenhouse. Material usually comes out of culture as well-started plantlets or microcuttings with a stockpile of nutrients and vigor often superior to that of conventional cuttings. It is no secret that healthy plants are the first line of defense against disease.

Most conventional nursery-grown seeds and cuttings must be grown during a particular season, and consequently, work schedules in the nursery must revolve around this factor. Although explant material may be limited to the season in which it can be taken, established tissue cultures can be grown at any time of year, regardless of weather. Working conditions in the tissue culture laboratory are ideal, and therefore conducive to year-round production, a situation that promotes maximum labor efficiency.

When using conventional propagation methods, one cutting produces one plant and one seed produces one seedling. In contrast, one explant—one piece of stem, leaf, bud, root, or seed, one meristem, or even one cell—theoretically can produce an infinite number of plants (Figure 1-3). Consequently, fewer stock plants are required to provide the explants needed to produce thousands of new plants. With the ever-increasing value of land and plants, more than ever the grower needs to put a monetary value on every bit of ground and every stock plant. Because tissue culture requires a minimal amount of plant material to start with, substantial savings can be realized not only by allocating less money to the acquisition and maintenance of stock plants, but also in the space and time required to maintain the stock plants.

In the greenhouse, cuttings can take months to root. The rapid multiplication typical of tissue culture often makes it possible to produce and sell true-to-form plants quicker and in greater numbers than one could by other means. The actual production time for tissue cultured plants varies depending on the particular plant. Whereas houseplants can start multiplying in culture within a matter of days or weeks, a woody plant can take several months before it begins to multiply. To give one conservative example, a single *Rhododendron* explant might take 4 to 6 months to start multiplying.

Once multiplication gets underway, assuming exponential multiplication (a doubling of material every month), from a single explant you would have 1024 plants after 10 months, 2048 plants after 11 months, and so on. Of course, you would hope to have more than one successful explant, and you should expect the explants to do more than double each time they are transferred to new cultures. In any case, while cultures are multiplying in such great numbers in the laboratory, greenhouse space can be used to grow other crops.

Tissue culture avoids an enormous amount of the daily care that is required with cuttings and seedlings. Cultures usually need to be divided and transferred to a fresh medium every 2 to 6 weeks, but between transfers there is no need to water or tend to the cultures, other than casual surveillance. How different this is from the daily watering and weeding requirements accompanying greenhouse growing!

For anyone who likes plants and is looking for a hobby, tissue culture can be a satisfying avocation. People can find great pleasure in watching and caring for plants in culture. It is an inviting field for the elderly or the handicapped because there are very few physical demands and the laboratory climate is generally comfortable. Retired folks, especially gardeners, who

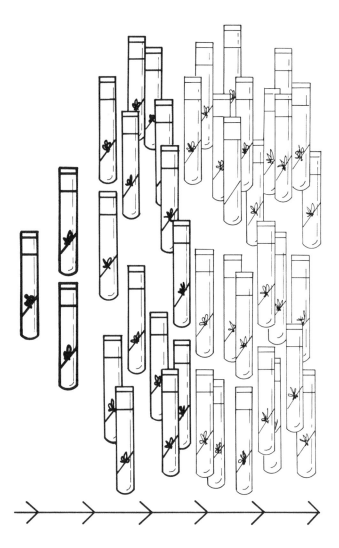

Figure 1-3. Tissue cultured clone. Potentially, one explant can produce an infinite number of plants.

want to take life a little easier can find many hours of interest and satisfaction in discovering how their favorite plant responds to tissue culture, all with a relatively small cash outlay.

Most larger laboratories involved in tissue culture have their own research departments to support the production operation. There are refinements of nutrients to be worked out, plants that have yet to be cultured, especially endangered species—to name just a few research objectives. The research department may be responsible for detecting and identifying contaminants in the production cultures, or it may be responsible for "fingerprinting," the identification of plants by separating genes or proteins in an electric field (electrophoresis).

The spectacular achievements that have been made behind the closed doors of elite laboratories do not prevent the curious minded from indulging in greater exploration, however. Hobbyists, amateurs not bound by the constraints of production, are free to pursue any avenue their aptitude, time, and budget will allow. Each will approach the field with unique background and insight, whether it be from academic or practical experience. Some will have studied chemistry, botany, or microbiology; others will have studied the ways of plants in the field, greenhouse, or garden. But all will have two characteristics in common: wonder and curiosity.

Much remains to be learned and studied in the field of plant tissue culture. Culture media are always subject to change and improvement. Techniques need to be improved upon and carefully described. Information needs to be made readily available to potential tissue culturists (except, of course, for priority information and patents, which is another subject). There are countless plants that have never been tissue cultured, many of which are in danger of extinction. What better legacy than to save some of these plants by means of tissue culture? These and many other areas are increasingly open to the amateur.

The simplest tissue culture hobby is the multiplication of easy, fast-growing plant material, such as *Kalanchoe*, Boston ferns *(Nephrolepis)*, African violets *(Saintpaulia)*, or *Begonia*; next in order of complexity are carnations *(Dianthus)*, strawberries *(Fragaria)*, or *Syngonium*. The chemist looking for a tissue culture hobby may be challenged to explore the field of plant by-products; dyes, flavorings, medicinals, and oils are just some of the by-products of certain plants.

The more studious (and affluent) hybridizer might be attracted to hybridizing plants by fusing *protoplasts* (cells without cell walls). Relatively few plants have been grown to maturity from single cells, but there is certainly enormous potential. The procedure requires more knowledge and equipment than simple micropropagation, but again, it need not be confined to the laboratories of academia.

Gardening is one of the most popular hobbies. Conventional gardening is limited by the seasons of the year, but tissue culture knows no season. Gardeners who propagate by tissue culture will delight in year-round micropropagation. If successful, they may find they have even more plants than anticipated. Excess plants can be shared with friends or offered for sale, and many ventures that start out as hobbies may turn into businesses.

The hobbyist or amateur gardener need not feel restricted in tissue culture pursuits for want of a transfer hood and a laboratory or any other fancy equipment, such as is required by commercial operations. Small-scale tissue culture is often carried out without benefit of a laboratory or special equipment. It can be done by almost anyone in almost any house. Material can be transferred on a desk or table in a clean, dust- and draft-free room of a home. Transferring cultures in a homemade chamber, with a glass or plexiglass front (see Figure 4-1) and just enough room for gloved hands to enter, is a reasonable method for the hobbyist. Furthermore, commercially available premixed culture media, plus a pressure cooker, forceps or tweezers, a paring knife, a few test tubes or jars, household bleach solution, and a lighted shelf, along with a lot of determination, are enough to bring about exciting discoveries for the amateur tissue culturist.

In contrast to the hobbyist, the commercial grower is compelled to make tissue culture a profitable enterprise. Growers with limited resources who must make a living from a small operation are finding that a large number of plants can be propagated by tissue culture with a minimal amount of space and outlay of capital. Oftentimes growers employ tissue culture techniques for one or two cultivars consistent with their operation, and then build a reputation for these cultured specialties. A few such plants that have made reputations for growers include carnations, ferns, iris, fruit-tree rootstock, orchids, and rhododendrons. In most cases, these growers used tissue culture propagation for plants to grow on to field-ready size; in other words, to develop rooted plantlets that will *grow-on* in the field to full-sized plants. Purchasing plants that are already started allows for a quicker return on capital for the commercial grower. On the other hand, in-house growing-on of plantlets from one's own laboratory is important to a nursery's normal market and precludes the competition that may be generated by selling the plantlets to other nurseries.

Growers, plant brokers, salespeople, greenhouse suppliers, and others interested in marketing plants have a variety of new products to offer: (1) plantlets still in culture; (2) plants directly out of culture, rooted or unrooted (microcuttings); and (3) plants hardened-off, or acclimatized, to greenhouse conditions. Most tissue cultured plants are true to type, more vigorous, more disease resistant, and disease free. When the end product costs less than conventionally propagated material, tissue cultured plants sell themselves.

Sometimes a grower welcomes the opportunity to buy rooted tissue cultured plantlets from other sources; in lieu of building their own laboratories, some small-scale growers with growing-on capability (i.e., who are able to take rooted plantlets and grow them to fuller size) will buy rooted plantlets from established commercial tissue culture laboratories, especially in the case of hard-to-start or -root or hard-to-find plants. There is room in the industry for the grower who will buy from a commercial laboratory plantlets just out of culture, and then harden them off to where he or she or other growers can finish and sell the plants. The wary tissue culturist, however, will guard against opportunists who might buy young tissue cultured plantlets, only to return the material to culture themselves, thus avoiding start-up costs while becoming an active competitor.

Before starting a laboratory, one should make sure there is a market for the proposed tissue cultured products, and then proceed gradually. It can be all too easy to buy quantities of equipment and supplies, only to discover that there is insufficient market demand for the product, or that the plant will not respond to culture, or that contamination is too difficult to remedy, or any other unforeseen problems. Many pitfalls can be avoided with a well-thought-out plan.

Pathogen-free plants from culture open the door to a freer exchange of plants between states and between countries. Plant tissue cultures have gained acceptance in world trade because the dangers of introducing disease is virtually eliminated. Foreign exchange will likely increase as new hybrids are developed asexually from protoplast fusion, and as other feats of genetic engineering find practical application. The free exchange of tissue cultures will have a significant impact on global food problems by allowing for more improved cultivars to be brought more rapidly to growers worldwide.

Although the application of tissue culture to farm crops has scarcely begun, important beginning efforts have been made with fruit trees, which are often difficult to propagate from cuttings. To maintain clonal characteristics, desirable scions (pieces of stem) are grafted onto special rootstocks, which are usually propagated by layering or cuttings (see Chapter 2). The rootstocks are cloned to maintain those that are known to influence certain aspects of scion growth, including such factors as hardiness, disease resistance, or dwarfing. Commercial tissue culture laboratories are currently culturing rootstocks with great success. If tissue cultured rootstocks prove economical, they will likely replace layered rootstocks. In fact, fruit trees are already being tissue cultured and grown on their own roots. If these efforts continue to be successful, grafting will come to be a cumbersome process of the past.

Whether the motivation for growing plants by tissue culture is profit, research, or personal satisfaction, the potential is there to produce a significantly greater number of healthier plants in less space, with less labor, and at less cost than by other means of vegetative propagation. The potential of plants is far greater than we know. It is as if the grower were a potter, little knowing what could be molded from the clay.

Chapter 2: The Botanical Basis for Tissue Culture

The remarkable diversity of naturally occurring vegetative reproduction reflects the amazing capability and potential of plants for multiplication. The same factors involved in multiplication and growth initiation in nature are involved in the greenhouse and in tissue culture. The natural capability of plants to multiply by asexual means is the basis for multiplication *in vitro*. No new phenomena have been invented for these processes, and genetic engineering and the manipulation of genes are not involved in tissue culture as such. There is no mixing of gene traits, as occurs in sexual reproduction. Vegetative reproduction, whether occurring naturally or through human intervention, is initiated in stems, buds, roots, or leaves.

Throughout history, careful observation of plant behavior and the study of plant anatomy and physiology have revealed a seemingly limitless amount of information, information that has taught us to manipulate certain natural phenomena to serve our own purposes. We take cuttings, we make divisions, we layer, we make grafts, and we tissue culture—in short, we promote vegetative reproduction, a natural phenomenon (Figure 2-1).

Plant stems have tremendous potential for regeneration. They grow in many different forms and habits: long or short, slender or stout, round, flat, or square, above ground or underground, trailing or upright. One of the methods of vegetative propagation used most often by growers is that of growing new plants from stem cuttings (Figure 2-2). Stem cuttings are shoots or sections of stems that root when they are inserted into a growing medium (potting mix), which may be a mixture of peat or bark with sand or perlite, or simply sand and perlite. Roots grow

Figure 2-1. A natural clone. Nature has been "cloning" for eons; any time a plant reproduces itself vegetatively it produces a clone.

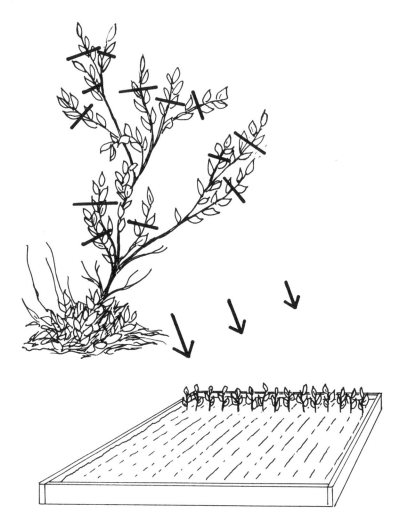

Figure 2-2. A grower-assisted clone by use of stem cuttings.
Growers have been cloning by cuttings for centuries.

from the *nodes* (the part of the stem from which buds or leaves originate) when placed below the surface of the medium. Frequently cuttings are encouraged to root by wounding, and often an *auxin* (a growth regulator) is applied to the base of the shoot. To *wound* a cutting, a thin slice of *epidermis* (outer layer of tissue) is peeled away with a knife on 1 or 2 sides of the lower part of the stem. This will often cause the stem to grow callus, which can help root initiation.

Small stem cuttings are frequently used as starting material for tissue cultures, and *microcuttings*, tiny cuttings taken from tissue cultured material, are a product of tissue culture that can be planted and grow on to mature plants. The size of microcuttings can be critical to the success of growth; 1 to 2 in (2.5 to 5 cm) is usually best.

Layering is another form of vegetative reproduction that occurs frequently in nature and is widely used by growers as well. A branch is said to layer when it comes in contact with the soil, forms roots, and grows on to become a new plant. Wild blackberries *(Rubus)* rapidly expand their territory by layering. Layering is used commercially to propagate filberts *(Corylus)*, grapes *(Vitis)*, black raspberries and trailing blackberries *(Rubus)*, currants *(Ribes)*, apple *(Malus)* rootstocks, and some ornamentals. *Air layering* is a method whereby growers induce a stem to root without it being in contact with the soil. The stem is wounded, auxin is applied, a layer of moss is wrapped around the area, and the whole is covered in plastic and tied. When roots are formed (a year or so later), the stem section can be cut away from the main plant and planted. The process of layering further demonstrates the ability of stems to root.

Nature can also be manipulated by growers in the process of grafting. In *grafting*, growers attach a *scion* (a shoot or bud) of one plant onto the *understock* or *rootstock* (rooted stem) of another plant to obtain the desired traits of both. The more closely the scion and rootstock are related, the greater the chance of a successful graft. A great deal of work and research concerning the compatibility or incompatibility of various plants has provided better understanding of which scions will graft successfully with which rootstocks. Several different methods can be employed in grafting, but whatever method is used, it is important that the cambial layers match—the *cambium* is the thin layer of tissue between the wood and the bark of a stem. *Micrografting* involves the grafting of tissue cultured material in aseptic conditions, a craft that requires great skill.

True bulbs, such as those of lilies *(Lilium)*, tulips *(Tulipa)*, and daffodils *(Narcissus)*, are underground storage organs that function as modified stems. A bulb consists of *scales* (modified leaves) attached to a *basal plate* (a flat modified stem at the base of the bulb and the source of roots). Bulbs multiply naturally by growing bulblets in the axils of their scales. Growers can encourage bulblet formation in some bulbs (such as *Narcissus* and *Scilla*) by cutting the bulbs from top to bottom, then further dividing them laterally. Each segment will produce bulblets if it contains a piece of the basal plate. Growers can promote bulblet formation in lilies by *scaling*, a process whereby the scales are removed from mature bulbs, dipped in a rooting hormone, and placed under growing conditions in which they will produce bulblets. Scoring and scooping are the means by which bulblet formation can be promoted in *Hyacinthus* and *Scilla*. *Scoring* is merely making cuts across the basal plate. *Scooping* is the process of cutting a cone-shaped piece out from the base. These methods induce the formation of more bulblets than would occur naturally without intervention.

Corms are swollen underground stems, complete with nodes, internodes (the section of stem between 2 nodes), and lateral buds; they do not have scales like true bulbs. Plants that produce corms, such as *Gladiolus* and *Crocus*, multiply naturally by producing *cormels* (miniature corms).

In addition to bulbs and corms, several other modified stem structures function in vegetative reproduction. Among these are *rhizomes*, horizontal underground stems, and *tubers*, the swollen fleshy part of a rhizome. Potatoes *(Solanum tuberosum)*, *Caladium*, and gloxinias *(Sinningia)* produce stem tubers. A piece of tuber or fleshy rhizome will generate a new plant if it contains an "eye," or bud. Other rhizomes are long and slender, such as those of lily-of-the-valley *(Convallaria)* and most perennial grasses. These rhizomes have long internodes with terminal and lateral buds, which allow the plants to multiply effectively. *Stolons* or *runners* are long, prostrate, above-ground modified stems. Examples of stoloniferous plants are creeping dogwood *(Cornus canadensis)*, strawberry *(Fragaria)*, Bermuda grass *(Cynodon dactylon)*, and white clover *(Trifolium repens)*. The sole purpose of stolons or runners is the production of new plants.

Roots take up water and nutrients from the soil and anchor plants firmly in the ground, but they also serve as useful structures for vegetative reproduction, both in nature and in commercial growing. Growers use root cuttings to propagate gooseberries *(Ribes)*, raspberries *(Rubus)*, horseradish *(Armoracia rusticana*, syn. *Cochlearia armoracia)*, apples *(Malus)*, flowering quince *(Chaenomeles)*, bayberries *(Myrica)*, aspens *(Populus)*, and roses *(Rosa)*, to name a few. The tubers of sweet potatoes *(Ipomoea batatas)*, *Dahlia*, and tuberous *Begonia* are modified roots; the buds formed on the stem end of these root tubers take nourishment from the food stored in the

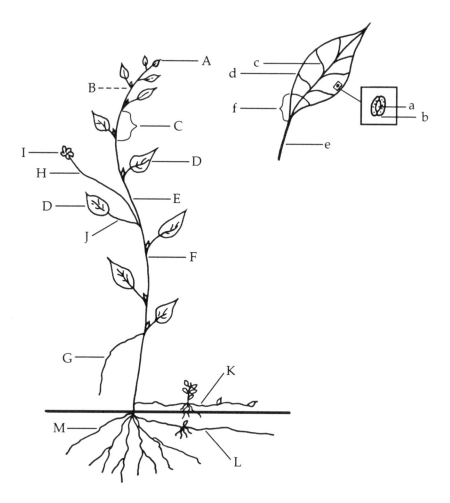

Figure 2-3. A hypothetical dicotyledonous plant: A. apex, apical bud; B. lateral (axillary) bud; C. internode; D. alternate leaf; E. stem; F. node; G. aerial, adventitious root; H. pedicel; I. flower; J. petiole; K. stolon; L. rhizome; M. root. Leaf: a. stomate (plural, stomata); b. guard cell; c. vein; d. hydathode; e. petiole; f. potential explant.

tubers as they grow into new plants. Tissue culture of roots is an important research tool for studying root development, mycorrhizae (beneficial fungi that associate with roots), and other soil organism–root interactions.

Leaves too will occasionally produce new plants. Leaf cuttings are made from such plants as *Bryophyllum*, *Begonia rex*, *Sedum*, African violets *(Saintpaulia)*, and *Sansevieria*, among others. In a few cases, new plants will grow from leaves without being separated from the parent plant; this happens in the piggy-back plant *(Tolmiea)* and walking fern *(Camptosorus)*. Leaves and parts of leaves are a common source of explants in tissue culture.

The foregoing examples from garden, field, and greenhouse illustrate the inherent potential, the power, and the inclination of plants for vegetative multiplication. These examples help to reveal the diverse anatomy of plants and some of their reproductive habits, and they serve as clues to areas and structures from which explants may be taken.

The multiplication of plants *in vitro* does not establish any new processes within the plants.

Tissue culture simply directs and assists the natural potential within the plant to put forth new growth and to multiply in a highly efficient and predictable way. In contrast to plants produced from seeds as a result of sexual reproduction, new plants produced through vegetative reproduction are basically severed extensions of the original plants.

New growth is usually initiated in *meristematic tissue*, which are undifferentiated cells that have not yet been programmed for their ultimate development. Meristematic cells are located at the tips of stems and roots, in leaf axils, in stems as cambium, on leaf margins, and in callus tissue. Under the influence of genetic make-up, location, light, temperature, nutrients, hormones, and probably many other factors, meristematic cells differentiate into leaves, stems, roots, and other organs and tissues in an organized fashion. Meristematic tissue is the basis of plant growth and development.

Parenchyma cells, the most common type of plant cell, are thin-walled cells that have the capacity to regenerate and *differentiate*, to initiate the growth of new and varied tissues or organs for specialized functions (Donnelly and Vidaver 1988). These and other cells that can *dedifferentiate*, or revert to an undifferentiated state, account for adventitious growth. *Adventitious growth* refers to the development of new shoots, buds, roots, or leaves from atypical and unusual locations. Examples of adventitious growth are aerial roots (roots that emerge from above-ground parts of the plant), buds from roots, plantlets from leaves, and shoots and roots from callus.

Dedifferentiated cells can also produce callus. Callus tissue can form in tissue cultures, it can form in response to wounding, or it can appear at the union of a graft or on the base of cuttings. The callus mass can contain *embryoids* (embryo-like somatic structures capable of developing into whole plants), or it can contain shoot or root *primordia* (the earliest developmental stage of an organ or cell). Callus can also develop cells with an abnormal number of chromosomes. For example, some *Asparagus* plants differentiating from a callus culture may be *tetraploid* ($4n$, or double the normal number [$2n$] of chromosomes in vegetative cells), but plants cultured from shoot tips, without a callus stage, usually do not show this variability. Callus culture dominated early tissue culture research and is a precursor for cell culture, but it is usually not necessary (and often is undesirable) for micropropagation.

There are 2 kinds of plant cell division: mitosis and meiosis. Every somatic cell is *diploid* ($2n$), containing 2 sets of chromosomes. During *mitosis* the chromosomes duplicate and then segregate (Figure 2-4). From this, 2 new cells form, each with chromosomes identical to those of the original cell. (See Chapter 13 for further discussion of these processes.)

Meiotic division, or reduction division, is the process of forming sexually reproductive cells. In *meiosis* each chromosome of a $2n$ cell splits in 2. The chromosomes segregate such that one chromosome from each set of 2 goes to each of the 2 new sex cells, or *gametes*, each of which has, thus, only one set of chromosomes (n).

Whenever cells divide there is the possibility of genetic variability. If a *mutation* (a change in genetic make-up) occurs during cell division, and assuming the cell survives the mutation, the mutation is carried in all future divisions. A mutation is a sudden abnormal change in genetic order that will alter some characteristic, or it can be a change in chromosome number. The effects of mutations are not always noticeable. Most mutations will cause a plant to die or to produce undesirable qualities, such as misshapen fruit or abnormal shoots. Growers expect a small percentage of plants to undergo mutation, and they will discard them if they see them. Fortunately, the natural frequency of gene mutation is very low—about one in a million cells for any single gene (Tortora et al. 1982). Occasionally, a mutation will create a desired effect, so scientists often induce mutations by treatment with chemicals or radiation in an effort to foster plants with improved characteristics.

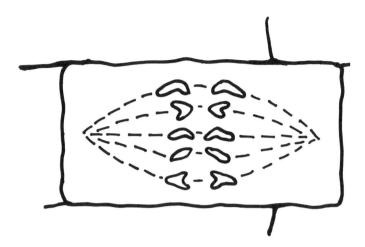

Figure 2-4. After the chromosomes duplicate, they separate before the cell divides.

Some variegated geraniums *(Pelargonium)*, *Sansevieria*, *Coleus*, *Croton*, and other variegated plants are chimeras. *Chimeras* are plants containing tissues of different genetic make-up growing adjacent to one another. Some thornless blackberries *(Rubus)* are chimeras, in which the aberrant characteristic (i.e., the lack of thorns) is limited to the epidermal tissue and can only be reproduced from stem tissues; new stems growing from roots will have thorns.

Navel oranges *(Citrus)*, which are seedless, and seedless grapes *(Vitis)* originated as *sports* (spontaneous mutations) and can only be reproduced vegetatively. Some abnormal divisions give rise to tetraploids (4*n*), as in the case of some giant apples *(Malus)*—a mutation that proved desirable. Such mutations, however, that are not desirable include the huge, unfruitful grapevines that appear in some vineyards. Any tissue cultures made from sections of variegated plants will exhibit only those characteristics of the plant from which the explant was taken.

The principles of tissue culture are all around us—in nature, in the field, and in the greenhouse. We learn from experience, from other growers, or from reading. We learn about the normal requirements for soil composition, nutrients, light, and temperature for a particular plant species. We study its form, its growth habits, and how it reproduces. These are only some of the clues that we can utilize in learning how to tissue culture plants. With this background information, we can turn to specific formulas and principles unique to tissue culture. But first we will review how these principles developed—a fascinating history, a story without end.

Chapter 3: History

To some degree the history of tissue culture involves the entire history of botany, the origin of which is lost in antiquity. A search through the years, however, finds the unique contributions of a certain few individuals especially relevant, and the events with which they were associated come together to provide us with an understanding of and appreciation for the past, as well as for the current state of the art.

Most early botany focused on the therapeutic value of plants, real or imagined. Throughout history, the merchants, missionaries, and crusaders who traveled near and far, returning to Europe with gold and fabrics, perfumes, herbs and spices, and many other goods from foreign cultures, also brought with them amazing stories of plant cures and remedies. With the advent of printing in the 15th century, knowledge and information began to spread as never before. In Europe, the medical traditions of Persia, of Egypt, and of Greece were studied and taught, focusing on the works of Hippocrates, Aristotle, Pliny, Galen, Avicenna (Ibn Sina), Theophrastus, and Dioscorides.

Dioscorides was one of the most important figures in this early history of botany. He traveled throughout the Middle East during the first century AD, observing and experimenting with plants. He made salves, extracted poisons, and found antidotes. He collected pistachio (*Pistacia*) resin and heated it for incense. He advocated the study of live plants as opposed to dried ones, as was the custom of most students at the time. He carefully described and recorded in detail the shapes, the growth habits, the fragrance, color, and beauty of the plants he observed. He also recorded in a lengthy work much of the botany that had preceded him.

The knowledge of Dioscorides and these other ancient scholars inspires awe, respect, and wonder even today. Without benefit of microscopes or modern chemistry, their studies relied on observation, reason, experience, tradition, and occasionally, magic and superstition. In their works can be found elements of today's botany: classification (taxonomy), form (morphology), dissection (anatomy), and function (physiology). There were also growers among these scholars—horticulturists, gardeners, and farmers. In all these individuals and all these fields lie the roots of plant tissue culture.

The microscope, an essential tool for examining and understanding cells and cell structure, is credited as having been invented by Zacharias Jansen, an eyeglass maker in Holland, in 1590. The following century, microscope lenses were significantly improved by Antoni van Leeuwenhoek, a Dutch merchant, linen draper, politician, and inventor, who turned his glass-blowing hobby to the making of lenses. Using a single-lens microscope, which he invented, Leeuwenhoek was the first to observe and describe one-celled microorganisms. He observed them in a drop of pond water using a microscope with a magnification of approximately 100× (Figure 3-1). What Leeuwenhoek observed was one of the most astonishing sights ever seen— the first view of spectacular microscopic life forms, the microbial world. From his careful drawings we know now that what he observed were protozoans, algae, and bacteria. He examined scrapings from his teeth, in which he observed microscopic forms darting back and forth through the water. In 1695 he recorded, "I then most always saw, with great wonder, that in said matter there were many very little animalcules very prettily a moving" (Robbins et al. 1965). His initial findings were submitted to the Royal Society of London in 1674. As a result of his pioneering work, Leeuwenhoek is known as "the father of microbiology."

A contemporary of Leeuwenhoek, Robert Hooke, an English physician, mathematician, and inventor, invented a compound microscope consisting of two lenses, an objective lens and an ocular lens (Figure 3-2). His findings also were reported to the Royal Society of London. He observed and illustrated the fruiting structures of molds and mosses, as well as parts from insects and small animals. Hooke was the first to apply the term "cells," and he identified cells as nature's building blocks of all living tissues, an idea that has come to be a cornerstone of biology. He counted—or perhaps more likely, calculated there to be—1259 cells in 1 cubic inch of cork.

A controversy surrounding how Leeuwenhoek's animalcules and Hooke's cells originate lasted for more than 150 years and had a devastating effect on the progress of biology. Some people argued that these microbes originated from other living cells, but others maintained that they generated spontaneously from nonliving matter (abiogenesis). Several early scientists tried to disprove the theory of spontaneous generation. Among them were Francisco Redi, who in the late 17th century proved that maggots came not from the decaying meat but from flies, and Lazzaro Spallanzani, who, nearly a century later, found that when broth was boiled and sealed in jars, microbes did not develop. Nevertheless, the controversy continued, and it was not until the mid-19th century that the theory of spontaneous generation was disproved.

In 1858 Rudolf Virchow declared that cells arise only from preexisting living cells—a theory that now seems so obvious—but people were not convinced until about 6 years later and the work of Louis Pasteur, a French chemist determined to improve France's wines. Before Pasteur people generally believed that air alone could introduce microorganisms; Pasteur's swan-

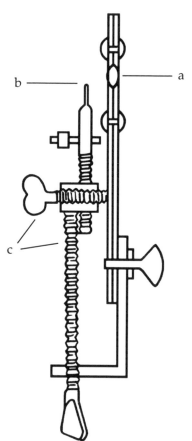

Figure 3-1. Model of a late-17th-century Leeuwenhoek microscope: a. lens; b. object being viewed; c. adjusting screws. The object being viewed is brought into focus with the adjusting screws. Very small in scale, these first hand-held microscopes were usually smaller than 6 inches (15 centimeters).

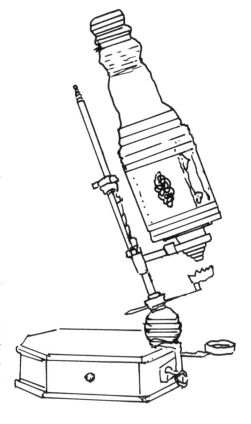

Figure 3-2. Early compound microscope of Hooke's time. Note the absence of a condenser lens.

necked flask experiment, however, proved to the contrary. Pasteur demonstrated that sterilized sugar or yeast broth could remain sterile indefinitely, even with air present in the flask, so long as no external bacteria or other microorganisms were introduced into the flask. In an effort to improve the flavor and lasting quality of France's wines, Pasteur developed a process of heating wine just enough to kill most of the bacteria—a process we now call pasteurization.

The experiments of John Tyndall, an Irish physicist and friend of Pasteur, reinforced Pasteur's conclusions and helped to finally silence those who claimed that unheated air, or the infusions themselves, contained a "vital force" that produced microscopic organisms. Demonstrating that different solutions or infusions require different lengths of boiling time for sterilization, Tyndall concluded that some cells were able to exist in 2 forms, one form that was sensitive to boiling and another that was resistant to boiling—these heat-resistant forms are now known as bacterial endospores (see Chapter 11). Tyndall showed that the endospores could be destroyed if the hay infusion was boiled at 3 different intervals. The first boiling killed all but the endospores. Following an incubation period in which the endospores could grow, a second boiling killed the germinating endospores. After another incubation period, a third boiling killed any late-germinating endospores. This process of sterilization is called *tyndallization*. Tyndallization is a tool useful for tissue culture in that it helps to determine the best method for sterilizing media for the growth of certain cultures. Through tyndallization, one can determine if it is preferable to use a boiled medium or a medium more conventionally sterilized in an *autoclave* (an enclosed chamber in which to sterilize equipment using steam to heat substances above their boiling points)—autoclaving can cause problems because certain chemicals will degrade or change under heat or pressure. For some of his experiments, Tyndall also constructed a chamber, the first recorded forerunner of present-day tissue culture *hoods*, which are boxes or chambers in which cultures are transferred aseptically (see Figure 4-1).

During the 1830s Matthias Jacob Schleiden, a botanist, and Theodor Schwann, a zoologist, further studied and speculated on the nature of cells. They observed that among lower plant forms a cell can be detached from the plant and continue to grow on its own. "We must, therefore," wrote Schwann (1839), "ascribe an independent life to the cell as such." Such was the beginning of the theory of cell *totipotence*, that any plant cell has the capability to regenerate the entire plant.

In spite of these inventions, theories, and discoveries of the mid-18th century, progress was slow, due in part to the use of microscopes that today would be considered primitive. In 1883 Ernst Abbe developed another important advance in the microscope. He greatly improved upon Zacharias Jansen's binocular microscope by adding a third lens, the condenser lens, located below the microscope stage (see Appendix B). By focusing this lens up or down it was possible to concentrate the light on the underside of the specimen, thus increasing image clarity. Abbe further improved image definition by introducing the use of lens immersion oil, which helps to deflect light rays into the lens, essential for magnification with the oil immersion lens (100×).

In any discussion of the history of the field of biology one must, of course, pay tribute to Charles Darwin, the king of all observers and author of *On the Origin of Species*, the famous work on the theory of evolution. In 1880 Charles and Francis Darwin first deduced the presence of a hormonal substance in grass *coleoptiles* (the sheath covering the seedling shoot tip). They observed grass stems bending toward light, yet when the tips were shaded the stems no longer bent. They proposed that some "influence" controlling the rate of growth flowed from the shoot tip to the growing region located some distance from the tip. Five years later E. Salkowski was the first to isolate this hormone found in plant shoot tips, but its growth-promoting properties were not fully recognized until years later.

English surgeon Joseph Lister provided the next important development in the field during the latter part of the 19th century. He applied the new theory of the time, which argued that certain microbes (germs) cause disease, to medicine. Using carbolic acid to clean surgical instruments, Lister was the first to use disinfectants. Lister's name is best remembered for the well-known mouthwash and antiseptic, Listerine.

Robert Koch, a German physician, further advanced the practice and understanding of bacteriology and *sterile technique*, the art of working with cultures in an environment free from microorganisms (see Chapter 8). Koch proved that specific organisms were responsible for causing specific diseases; specifically, anthrax is caused by *Bacillus anthracis*, and tuberculosis is caused by *Mycobacterium tuberculosis*.

Justus von Liebig, a German physical chemist, theorized in 1840 that the minerals found in soil were essential to plant growth. His particular contribution was the concept of limiting factors: if any one essential plant nutrient is missing, all the others are of little benefit. Twenty-five years later, Johann Knop developed a nutrient solution based on soil analysis. Knop's solution was used by early experimenters in soil-less culture, and it is still used today (Knop 1865).

In 1902 the German botanist Gottlieb Haberlandt used Knop's medium, supplemented with sucrose, asparagine (an amino acid), and peptone, to grow cells, but the cells lived only a few weeks. Haberlandt also predicted that plant embryos could be produced by cultivating vegetative cells. Two years later E. Hannig, another German botanist, anticipating the procedure of embryo rescue, successfully cultured premature, excised crucifer embryos. He observed the embryos produce small weak plantlets in culture, instead of developing into normal embryos. He called this "precocious germination."

As the 20th century progressed, the field of plant tissue culture embarked on an era of exponential growth. Of early commercial significance was the germination in aseptic culture of orchid seeds and seedling growth on a nutrient medium with *agar*, a polysaccharide gel derived from certain algae. This feat was reported independently but almost simultaneously by L. Knudson, Noel Bernard, and H. Burgeff in the early 1920s. At about the same time, W. Kotte, a student of Haberlandt's, and William J. Robbins independently accomplished limited success in the culture of root tips.

Early experiments confirmed that tissue cultured plants while in culture were *heterotrophic* (unable to manufacture their own food from inorganic substances), as opposed to *autotrophic* (capable of manufacturing their own food). Unlike plants in soil, plants in culture cannot manufacture proteins and carbohydrates from inorganic nutrients. Perhaps empirically, it was discovered that sugar and undefined substances such as coconut milk, yeasts, and fruit juices were able to support culture growth, but inorganic chemicals alone could not. Robbins reported that tomato *(Lycopersicon)* root tip cultures were aided by the addition of a yeast *(Saccharomyces)* to the medium. Later analysis revealed that yeasts contain certain desirable vitamins, particularly thiamine (vitamin B_1).

Callus culture of carrots *(Daucus carota)*, a classic subject of investigation, was reported by two physicians, R. Blumenthal and P. Meyer, in 1924. They were primarily interested in the pathological implications of such a study, comparing callus to tumor growth, but in 1927 L. Rehwald demonstrated the cultivation of callus from carrot slices, irrespective of the pathological considerations.

Drawing on Darwin's earlier observations of coleoptiles and on the work of Salkowski, in 1911 P. Boysen-Jensen demonstrated that the growth-promoting substance found in plant shoot tips would diffuse across a wound covered with gelatin, and he realized that the substance could diffuse out of the tissue and into a collecting gel. In 1928 Frits Went collected this growth substance from coleoptile tips in tiny blocks of agar. When he placed an agar block containing the substance asymmetrically on top of a decapitated shoot, the shoot proceeded to grow more on the side with the block than on the side without the agar block. The greater the number of tips that were used for collecting the substance in a particular agar block, the greater the curvature in the growth of the shoot. In 1934 F. Kogl, A. J. Haagen-Smit, and H. Erxleben were the first to isolate and chemically analyze the substance. They identified it as a plant hormone, or auxin, which they named indole-3-acetic acid (IAA). Five years later, R. J. Gautheret and P. Nobecourt, both in France, independently reported indefinite growth of callus from carrot cambium when using auxin in the nutrient medium. Gautheret had been the first to successfully culture plant tissue. In 1934 he grew cambial tissue of sycamores *(Acer pseudoplatanus)* and, later, of pussy willows *(Salix caprea)* and elders *(Sambucus)*, some of which continued in culture for more than a year.

P. R. White is acknowledged as the father of tissue culture in the United States. He was the first to grow excised root tips of tomatoes *(Lycopersicon)* in continuous culture. The inclusion of glycine (an amino acid), pyridoxine (vitamin B_6), and nicotinic acid (niacin, vitamin B_3) in his root tip media were important and effective additions to the experiment. In 1939 he reported successful culture of tobacco *(Nicotiana)* callus. In collaboration with Armin Braun, White demonstrated the similarity of tumor cells in plants and animals by growing tumor tissue from tobacco. White's *A Handbook of Plant Tissue Culture* (1943) brought together the accumulated knowledge of plant tissue culture as it stood at the time.

Many scientists in the mid-20th century were working on culturing embryos extracted from seeds *(embryo rescue)* or attempting to stimulate the spontaneous production of embryos from undifferentiated cells *(embryogenesis)*. In light of this current state of the research, a logical next step seemed to be to investigate the use of coconut milk, the liquid *endosperm* (nutritive liquid or tissue of seeds) from coconuts *(Cocos)*—a ready-made, natural nutrient medium within seeds that nourishes the embryo. Coconut milk was first used in 1941 by J. van Overbeek, M. E. Conklin, and Albert F. Blakeslee, who discovered that it stimulated callus formation in cultures of excised embryos of jimson weed *(Datura stramonium)*. F. C. Steward, a renowned plant physiologist at Cornell University in New York, was so impressed by the dramatic effects of coconut milk in carrot culture media that he set aside his other objectives in order to dedicate himself to the study of growth factors in this and other liquid endosperms. Among the active materials he extracted from the coconut milk were several ingredients that are now commonly included in purified form in many tissue culture media. Coconut milk is still used in some orchid culture media.

The orchid industry was the first to apply micropropagation on a commercial level. George Morel and C. Martin cultured virus-free *Dahlia* shoots and potato *(Solanum tuberosum)* plants by meristem culture, following protocols set forth by E. Ball in 1946. In 1960 Morel and Martin applied their findings to orchids. Not only did they succeed in freeing the orchids of viruses, but unexpectedly, they also observed multiplication of their cultures, a phenomenon (micropropagation) that would in time revolutionize many features of the entire horticultural industry. Orchids, for example, have become more plentiful and less expensive as a result of this discovery, and many tested disease-free cultivars are now readily available because of micropropagation.

In spite of these early accomplishments, the need to further define media ingredients and their proper proportions for successful commercial application remained an important issue. Several discoveries during the 1950s addressed such concerns. In 1955 C. O. Miller discovered *kinetin*, a hormone that promotes bud formation and the first of a group of plant growth regulators known as *cytokinins*. Frits Went collaborated with Folke Skoog to examine the bud-inhibiting effects of auxin and its interaction with kinetin. Went and Kenneth V. Thimann demonstrated the root-initiating properties of the auxin IAA. Later, in 1957, Skoog and Miller published "Chemical Regulation of Growth and Organ Formation in Plant Tissues Cultured *In Vitro*," in which they discussed the desired auxin/cytokinin ratios for the growth of cultures.

Skoog's name is immediately recognized by anyone involved in tissue culture for his part in the Murashige and Skoog medium formula, commonly referred to as M & S or MS medium. The medium, now used almost universally, was reported in the classic 1962 article "A Revised Medium for Rapid Growth and Bioassays with Tobacco Tissue Cultures." Higher in salts than previous media, with additional subsequent modifications, the MS medium provided a magical key to culturing many more plants than had previously been possible. Elfriede Linsmaier collaborated with Skoog shortly thereafter with a report of their systematic study of the organic requirements of tobacco *(Nicotiana)* callus. They indicated some appropriate adjustments of the MS formula, and the groundwork was laid for an even wider application of this versatile formula.

Toshio Murashige, a former student of Skoog's at the University of Wisconsin and later a professor at the University of California at Riverside, did much to bridge the gap between research and industrial applications.

NEW TECHNOLOGIES

In the broad field of the natural sciences, there is an increasing recognition of the interrelatedness and interdependency of the many particular fields and disciplines; physics, chemistry, botany, zoology, algology, mycology, genetics, and others, can no longer be isolated and departmentalized, but must be studied in context with one another. In light of this, new developments in plant research inspire the use of tissue culture in several overlapping areas. Some of these that are of broad interest include anther, callus, cell, and embryo culture, protoplast fusion, mycology, algology, secondary products, and genetic engineering. See Chapter 13 for further discussion of some of these applications.

Liquid cell suspension cultures have particular significance for the mass production of cells. One common source of cells for cell suspension is from friable callus, although specific cells, such as from leaf *mesophyll* (the thin, soft tissue between the upper and lower epidermis of the leaf), are also grown in suspension. Cells in suspension can form embryoids (somatic embryos) in the process of embryogenesis. Embryos may multiply and/or be induced to form plantlets in the process of *morphogenesis*. Many hybrid plants produce embryos that do not mature to viable seeds. These embryos can be "rescued," removed from the seed in an immature stage, and then grown in culture.

Suspension cultures have been enhanced by new methods that can continuously introduce fresh medium into the suspension culture, thereby enabling the production of thousands of cells or embryos in a single container with a minimum of manual transfer. This is one way that tissue culture can compete with the plentiful seed production in nature.

Interest in anther and pollen culture—the tissue culturing of anthers or pollen to obtain *haploid* (cells with half the normal number of chromosomes of vegetative cells) clones—is spurred by the practical applications of such haploid cultures. Haploid *(n)* plants are sterile, but if the chromosomes duplicate, either spontaneously or by induction, the plants will be diploid *(2n*, which is normal for the vegetative state), and their progeny will be true to form. Considering the fact that it takes several generations of inbreeding to obtain a pure line by conventional means, it is little wonder that plant breeders are interested in anther culture.

Genetic engineering has also brought about several new useful products and processes. The 3 most common ways of altering the genetic make-up of plants are protoplast fusion; *muta-*

genesis (the production of cell mutations) using chemicals or radiation; and genetic engineering whereby plants are transformed by the introduction of foreign genes.

When the cell wall of a cell is dissolved away, the remaining membrane and its contents comprise a protoplast (Figure 3-3). Because they lack cell walls, protoplasts are more permeable and, as such, are particularly useful for studying the intake of or resistance to toxins, nuclear material, viruses, bacteria, or fungi. Of particular interest to plant breeders is the fact that protoplasts will fuse with other protoplasts not only of the same clone, but also with those of other species or genera. Before 1960, cell walls had to be removed by physical means (microsurgery). Cells were plasmolyzed by placing them in a solution that caused the protoplast to shrink away from the wall. The cells were then randomly cut up with a sharp scalpel, and the undamaged protoplasts were recovered with a pipet. In England in 1960 E. C. Cocking developed and published a method for removing cell walls by chemical (enzymatic) methods. Complete plants have been regenerated from protoplast cultures in at least 30 plants, including tobacco *(Nicotiana)*, wheat *(Triticum)*, carrot *(Daucus carota)*, *Asparagus*, potato *(Solanum tuberosum)*, and tomato *(Lycopersicon)*.

After many failed attempts by numerous experimenters to induce fusion between protoplasts from different sources, in 1974 K. N. Kao and M. R. Michayluk discovered that when a solution of polyethylene glycol is added to a mixed protoplast culture, followed by high calcium dilution, a significant percentage of protoplasts will fuse. It was soon discovered that electrical impulses are another inducing agent. With these methods, protoplasts of tomato plants have been fused with protoplasts of potato plants, and the resulting product was able to differentiate into complete hybrid plants. The objective was not to create a new "freak," but rather to convey certain disease-resistant qualities of one plant to another by means of an intermediary plant. So far, only very few fusions between different species have developed into

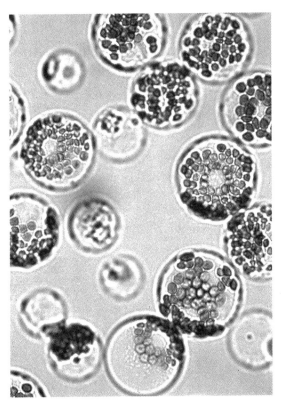

Figure 3-3. Unstained potato protoplasts. In order to prevent disruption, protoplasts are often kept in a sugar solution, such as mannitol, which has the same osmotic pressure as the protoplast cytoplasm. Courtesy Dr. Sandra Austen, Department of Biotechnology, University of Wisconsin.

mature plants, but the potential is there to produce plants with greater food value, more resistance to pests, and better ornamental qualities.

Mutations in plants occur naturally at a very low rate (about one cell in a million). They can be induced by artificial means, however, and tissue culture offers a convenient method of handling plant material when it is treated with *mutagens* (mutant-inducing agents) such as radiation, ultraviolet light, or carcinogenic chemicals. Following treatment, cultures are incubated and then tested for certain characteristics, such as resistance to toxins, salts, herbicides, antibiotics, or disease. The cultures can also be tested for their tolerance to heat or cold, their hormone and nutrient requirements, or their ability to synthesize secondary products, or *metabolites* (products of *metabolism*, the life processes of all living matter).

The production of secondary products is a particularly promising application of plant tissue culture, and personnel from oil- and food-producing companies are prominent among students of tissue culture. Some cell cultures can be induced or genetically engineered under controlled conditions to produce useful compounds in greater quantity and at less cost than can be obtained from whole plants. A sampling of the types of secondary products potentially available from cell cultures includes flavorings, pigments, medicinals, gums, resins, antibiotics, insecticides, fungicides, alkaloids, enzymes, and oils. The products are extracted chemically and physically either from the cells or from the medium, depending on whether or not they are released from the cells.

The origin of genetic engineering in plants can be traced to the observation of the condition of *hyperplasia*, an abnormal increase in cell number causing irregular swelling or growth, which is often a response to disease. The tumors caused by crown gall *(Agrobacterium tumefaciens)* are a familiar example of cells that have undergone natural genetic transformation (see Figure 13-5).

Because it is possible to determine the specific gene or genes responsible for certain plant characteristics, a genetic engineer can insert into a plant the genes that code for a desired trait or traits. These genes can be obtained from another plant, from a bacterium, or even from animal cells. They may code for cold hardiness, insect resistance, disease resistance, proteins, enzymes, medicinals, flower color, and many other qualities. The importance of this research must not be underestimated.

In the 1980s Kary B. Mullis invented PCR (polymerase chain reaction), for which he was awarded a Nobel Prize. This outstanding system is a process that automates the exponential multiplication of *DNA* (deoxyribonucleic acid), a carrier of genetic information. The double-stranded fragments are denatured (broken down) by high temperature and are dissociated to single strands. Each new piece serves as a template for another new copy. *Polymerase*, the *enzyme* (a specialized protein that acts as an organic catalyst in specific chemical reactions) used in the duplication part of this process, was extracted from bacteria from hot springs. Because the enzyme can withstand the high temperature that is required for the denaturation part of the process, it is a key player in the duplication of single-strand DNA.

This brief summary of some of the most notable recent activities involving tissue culture should stimulate the imagination and alert the commercial propagator or student to the broad opportunities offered by the new technologies. The tangible products of these research efforts will come as no surprise to the knowledgeable propagator. They appear in the nursery trade, in the corn and wheat fields, in the kitchen, in the pharmacy, in the flower shop, and in the bowls of the hungry.

Chapter 4: The Laboratory— Facilities and Supplies

Any person considering a tissue culture operation will approach the project with a unique background, with different goals, pressures, and resources. Before planning a laboratory, you should determine the magnitude of the operation, the purpose or goals, and to some extent, the plants to be cultured. How elaborate a facility do you want or can you afford?

A hobbyist can reap great pleasure, and a little pocket money, from a small laboratory that serves personal or local interests. At the other extreme—a large laboratory serving the domestic market or shipping to foreign countries—the potential market and the distance from consumers are major concerns. A preliminary assessment of potential markets is an important early step; find out in advance who will buy your product and at what price. It is often advisable to establish a pilot plant operation in order to asses the feasibility of producing the plants of your choice.

If possible, visit government, school, or commercial plant tissue culture facilities. Observing these large-scale facilities will help you to formulate your own plans and can serve as models. Also visit nurseries and florists, both wholesale and retail, to discover what plants are in demand or hard to get. Visit government agents familiar with horticultural businesses to learn about the current needs and problems of the plant industry. Learn from government agents or exporters what are the current regulations for shipping from state to state or to foreign countries. At one time tissue cultured plantlets moved freely through customs; but now it is understood that even cultures which appear healthy can harbor viruses or other unsuspected pathogenic, systemic, microbial contaminants that may not become evident until much later, when the plants are growing in the field or garden. Consequently, regulations have become more strict and vary with state or national governments.

LOCATION

Individual circumstances will, of course, determine the location of a proposed laboratory, but some general concerns need to be addressed in any location. Are there existing buildings and greenhouses that can be converted to a laboratory and growing-on facilities? In all aspects of planning, cleanliness should be considered one of the most important factors. Air quality and direction of air flow are of particular concern because unsterilized air carries microbial contaminants. Select an area that is relatively free from dust, smoke, molds, spores, and chemicals. Driveways and parking areas should be close enough for convenience, but far enough away that blowing dust will not affect the facility. A well-kept lawn and disease-free shrubs will not only contribute to an attractive entrance, but will also promote a clean atmosphere. Schools should locate a plant tissue culture laboratory as far away as possible from a plant pathology laboratory.

A hobbyist may start tissue culturing in a home kitchen, using a homemade transfer chamber (Figure 4-1) and a bookcase with lighted shelves. This setup will serve quite well for a while, but the serious amateur who masters the techniques and achieves rewarding results in limited facilities will soon want to advance to a more sophisticated laboratory for pleasure and profit.

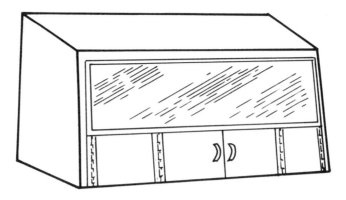

Figure 4-1. A still-air box (transfer chamber) for the hobbyist.

DESIGN

A plant tissue culture laboratory consists of 3 distinct areas: (1) a space for preparing media; (2) a chamber for transferring cultures; and (3) an area in which to grow the cultures. Usually these are located in 3 separate rooms, but when just starting out or in the face of other limiting circumstances, these areas may occupy the same room.

The media preparation area resembles a kitchen; there should be a sink with hot and cold running water, a refrigerator, a dishwasher, and a stove or burner. You will need a pressure cooker or an autoclave for sterilization. Counter space should be made available for a balance, a pH meter, and a hot plate/stirrer, as well as room to prepare the media. There should be a place for water treatment and storage. Cupboard space is required for storing glassware, chemicals, and other supplies. The preparation of media requires several steps, which the space must accommodate. Chemicals must be weighed and added to water. A medium containing agar must be stirred while it is heating so that the agar will melt. After the agar has melted and the chemicals are dissolved, the medium must be dispensed, sterilized, and stored, ready for use.

The transfer room contains the transfer chamber or hood (see Figure 4-12), which is usually a bench with a partially enclosed area that is provided with sterile air. It is here that the technician starts, divides, and trims the cultures and transfers them from one container to other containers (see Figure 4-13). Between transfers the cultures stay in the warm, clean growing area, where well-lighted shelves hold the vessels (test tubes, bottles, flasks, or petri dishes) in which the cultures are growing.

The transfer room and the culture growing room should be isolated as much as possible from outside doors and from significant foot traffic. Interior pass-through windows between rooms can reduce traffic, help prevent contamination, and minimize the opening of doors, but they are not practical for larger operations, in which case large numbers of containers should be moved on carts. The use of tables or carts is an effective way of increasing counter space in any of the 3 areas. The size of cart can vary from that of a tea cart to a bank of shelves on casters. All doorways should be wide enough to accommodate carts.

Windows may be placed conveniently in the media preparation room and, sometimes, in the transfer room, but never in the culture growing room because outdoor light and temperatures reduce the ability to control the growing environment. One objection to windows in the transfer room is that the irregular surfaces on window frames can be difficult to clean properly;

on the other hand, many workers find it a more comfortable environment if they can see out through a window. Light-colored walls and music also can help create a more comfortable working environment. Studies have shown that work efficiency drops off after 6 hours, so the best possible environment should be built into the transfer room. In any case, if there are windows in the transfer room, they must be tightly sealed.

In all rooms, the walls, floors, and ceilings should be light colored and easily washed with detergent and standard disinfectants. Banks of shelves should be set high enough off the floor to allow for convenient cleaning of the floor underneath and to help avoid contamination from the floor.

Any outside air that is admitted into the 3 major rooms should be filtered, preferably by *HEPA* (high efficiency particulate air) filters, in order to reduce the particle count to an acceptable level. The particle count for the transfer room and growth rooms should be 10,000 particles or less per cubic meter of air, whereas 100,000 is adequate for the media preparation room and research areas. Rooms used for such functions as office and shipping do not require filtration of room air.

Power and wiring

Planning for electrical equipment and fire safety always deserves the best available professional advice. This is especially true for a tissue culture facility because of its high demand for electrical equipment and, in some instances, the use of open flame. Most of the wiring can be for 110 volts, but heating, cooling, water treatment equipment, and autoclaves often require 220 volts.

Heating and cooling can be supplied by electric wall heaters or furnaces and air conditioners, but a heat-pump system is far better and well worth the initial expense. A heat pump should contain appropriate filters, such as an in-line electrostatic air filter and, possibly, HEPA filters for transfer and growing rooms. Information regarding developments in heating, lighting, and energy conservation can be obtained from your power company and from heating and lighting equipment manufacturers. It is wise to confer with them before making final decisions about equipment choices, which need to satisfy the plants as well as the workers who must occupy the same environments; for example, cool-white light is desirable for plants, but daylight-quality fluorescent fixtures may be preferable for people.

Electrical power failure and safety are other major concerns. It may be necessary to obtain a portable generator to keep the growing room at the correct temperature in the event of a power failure. Safety plans should include built-in escape routes and the location of firefighting equipment.

Growing room lighting and temperatures

A typical growing room uses cool-white fluorescent lamps above the cultures. Two 8-ft (2.4-m) cool-white fluorescent fixtures, spaced 20 in (50 cm) apart, provide 100 to 300 foot-candles (1000 to 3000 lux) of light to the cultures on each growth shelf. The academic literature often describes light intensity in different units, suggesting ranges for tissue cultures from 50 to 1200 micromoles per square meter per second (μmol m^{-2} sec^{-1}) (Preece and Sutter 1991), or from 20 to 100 microeinsteins per square meter per second (μE m^{-2} sec^{-1}) (Smith 1992). Light meters reading in foot-candles (f.c.) and lux are readily available and still serve their purpose for most tissue culture laboratories. Fortunately, cultures appear to have a wide tolerance for light intensity.

With 2 or even 4 fluorescent tubes over a 4-ft (1.2-m) wide shelf, it is not possible to have uniform lighting reach all of the cultures. The light reaching the cultures varies with the distance from the light source, the type and size of culture container, and the density and size of the cultures. On shelves separated by about 16 to 18 in (40 to 45 cm), the fluorescent tube will be about 10 in (25 cm) above the shelf, taking into account the depth of the fixture. The light intensity on the cultures directly below the lamp might be 3 times that reaching the cultures at the edge of the shelf.

The spectral curve of light from Gro-Lux lamps appears to coincide more accurately with the requirements of chlorophyll synthesis than does cool-white light (see Chart 4-1); however, common practice and some comparison studies indicate that cool-white fluorescent tubes are adequate and even preferable for most cultures. Occasionally Gro-Lux lamps are specified in the literature for a specific plant, in which case one can assume that the author's preference was justified. If you are considering using these lamps, first test their effectiveness and safety on a few samples because they have been known to burn some plants. General Electric has replaced what has been known as cool-white with F96P12.

More-expensive fluorescent fixtures can modify or eliminate the problem of heat given off by ballasts. The temperatures of hot spots in the shelves directly over the ballasts can become so hot that it will cook the cultures; therefore, the ballasts are often removed and placed at the ends of the shelves in a protective metal box or installed outside of the room. Some solid state

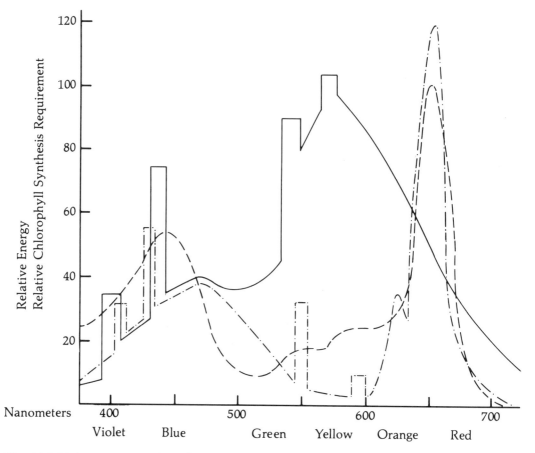

Chart 4-1. Light wave comparison chart.
Chlorophyll synthesis   – – – – – –
Cool-white radiance   ———
Gro-Lux radiance   – · – · – · – ·

ballasts offer the advantages of less heat, no flicker, and no noise. Consult your local lighting or electrical authorities for the latest technical information.

Culture growing room temperatures generally should be kept between 75 and 85°F (24–29°C), with 16 hours of light and 8 hours of darkness, although this will vary with the particular plants being grown. A centrally located thermostat controls heating and cooling equipment. Such equipment usually has a blower that can be left on continuously to provide good air circulation, which is essential for minimizing hot areas over lights and ballasts, or cold areas near the floor. Sometimes shelves are equipped with miniature inflatable air ducts to help circulate the air.

A 24-hour timer can be used to automatically control the lighting and help balance the heating and cooling demands of the laboratory. The timer should be set so that the 8-hour period of darkness occurs during the daytime and the 16 hours of light fall mostly during the night. By this arrangement, the heat given off by the lights at night helps keep the growing room warm when heat is most needed and allows the room to be cooler in the daytime, to save cooling costs. Usually a culture's light-cycle requirements will not be noticeably affected if you turn on growing room lights periodically during the day for doing chores. A light meter, reading in foot-candles, should also be purchased to measure light distribution and intensity; they cost about US $90 (Edmund Scientific F34,178).

Examples of laboratory designs

A simple plan for an operation incorporating some of these design features is shown in Figure 4-2 (the laboratory that is illustrated was the original lab of Lydiane Kyte). The plan lends itself to a variety of options, and the total area is adequate for culturing a million plants. The laboratory was built inside an existing metal building. The media preparation room is 16×12 ft (4.8 $\times 3.6$ m), the transfer room is 8×12 ft (2.4 $\times 3.6$ m), and the culture growing room is 18×12 ft (5.4 $\times 3.6$ m). The transfer chamber can be widened to accommodate 2 work stations, allowing 2 technicians to transfer at the same time. If adjacent counter space is limited, carts are a good supplement, especially for finished test tubes or jars ready to be moved to the growing room. The transfer room can also be expanded into the adjacent entry area. Either entry area can be developed further for storage, office, autoclave, water treatment, or rest rooms. Cupboards above the bench or counter in the media preparation room are desirable, but they may not be required initially.

The laboratory of Briggs Nursery in Olympia, Washington, also is based on a well-designed plan (Figure 4-3). The operation produces several million plantlets a year with only about 10 employees. It is in a building constructed within a metal building which is 80×60 ft (24 $\times 18$ m). Most laboratories are not of this magnitude, of course, but the plan can be scaled down and the concepts and product flow are worthy of consideration.

The transfer room, growing room, and research areas are under positive air pressure. The ballasts for the fluorescent lights in the growing room are located in the attic area, reducing the excess heat in the growing area and also limiting any direct hazards from the ballasts, which have been known to explode. The clean-up room is where contaminated cultures are autoclaved and dishes are washed; it is also the area from which plantlets are sent to the greenhouse and to which the containers are returned for cleaning and re-use. Four counter-top autoclaves are located in the clean-up room and the media preparation room.

The office also serves as a conference room and library. The hallway along 3 sides of the laboratory provides escape, access, and insulation. Sliding glass doors provide access and visibility between the transfer room and media preparation room, between the transfer room and growing room, and between the office and research room.

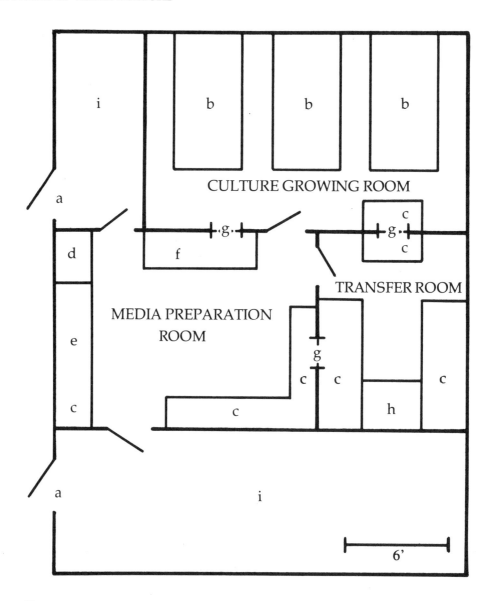

Figure 4-2. A small tissue culture laboratory: a. entry; b. shelves; c. counter; d. refrigerator; e. sink; f. stove-top burner; g. pass-through window; h. transfer chamber; i. space for future expansion.

Visitors to the Briggs Nursery laboratory are ushered from the main entrance, past the locker area, to windows through which they can view the transfer room and the growing room. In the middle of the transfer room are the 8 6-ft (1.8-m) transfer hoods placed back-to-back.

The laboratory of Micro Plantae Ltd., in Pune, India, (completed in 1993) has some innovations worthy of note. They have a production capability of 10 million plantlets per year, using 2 shifts a day with a work force of 20 people per shift in the transfer room. Two autoclaves, each 4 × 6 ft (1.2 × 1.8 m), are used for media sterilization. The autoclaves can be opened at both ends, thereby enabling removal of sterilized media directly into the transfer room storage area. In so doing, they avoid diminishing the air quality and cleanliness in the transfer room. The glass bead sterilizers (see Figure 4-10) for transfer tools are mounted below the level of the sur-

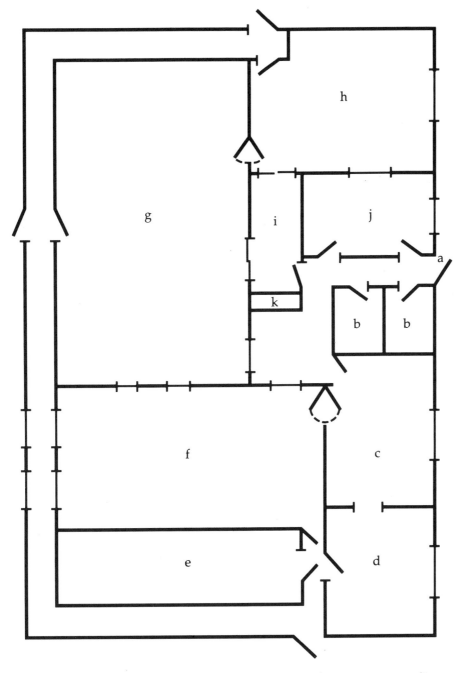

Figure 4-3. A large tissue culture laboratory: a. main entry; b. restrooms; c. media preparation room; d. storage/clean-up; e. cooler; f. transfer room; g. culture growing room; h. research room; i. office; j. conference room; k. lockers. Courtesy Briggs Nursery, Olympia, Washington.

face of the transfer hood, again helping to maintain a sterile working surface. Each hood contains an ultraviolet light unit that is turned on for 30 minutes before each shift to sterilize the bench surface. (Ultraviolet lights are rarely recommended due to the health hazard they pose to eyes and skin.) Carts are used to move trays containing small culture jars to various production areas. When leaving either the transfer or growing rooms, the carts are moved through a pressurized interlock (double door) area.

Attached to the transfer room at the Micro Plantae laboratory is an office that has a computer for entering production data, such as the location of each batch of culture jars in the growing room. In an operation of this magnitude, such record keeping is essential.

The growing rooms measure approximately $30 \times 30 \times 20$ ft ($9 \times 9 \times 6$ m). The shelves extend to the ceiling, necessitating the use of ladders to reach the higher shelves. By using wire-mesh trays and shelves, plus an excellent air-circulation system, the laboratory is able to regulate air temperatures throughout the growing rooms to within about 1 or 2°F (1°C).

The above factors, together with the maintenance of positive air pressure in both the transfer room and the growing room, allow for an air cleanliness standard in the Micro Plantae laboratory of 10,000 particles per cubic meter of air (as detected by an active quality-control program). No doubt an in-depth operator training program also helps Micro Plantae keep the number of contaminated containers at a level of 1% or less.

Do not be discouraged. If you have a new laboratory, with perhaps only a laminar air flow hood (transfer chamber) and a high rate of contamination, such measures at first may seem overwhelming. Rest assured, your contamination levels will decrease as you become more familiar with quality-control procedures.

EQUIPMENT

Water purification equipment

Water is the largest component of tissue culture media, so the quality of the water used is of critical importance in establishing a successful operation. Tap water often contains dissolved minerals, particulates, and organic matter. These substances must be removed before using the water in tissue culture media because they can upset the precise balance of nutrients in media formulas, or they can be toxic. Any imbalance can promote precipitates (insoluble chemicals), which make the nutrients in media unavailable to the plantlets.

A convenient way to obtain pure water is to buy bottled distilled water. Both deionized and distilled water are available at local supermarkets in gallon jugs, usually for less than US $1.00 per gallon (3.8 liters). It is advisable to buy bottled purified water when starting on a small scale and limited budget, but for a permanent operation a still or deionizer should be purchased.

Rain water can also be used, but it must be carefully tested, particularly with regard to the means of collection. Rain water can pick up undesirable elements from a metal roof, or it can grow undetected organisms that give off undesirable organic compounds, either of which might not be eliminated even with subsequent treatment.

One common method of ascertaining the purity of water is to measure its conductivity. *Conductivity* is a measure of the ability of water to conduct electricity. An electrical current passes through water by the movement of ions, so the conductivity level indicates the concentration of ions dissolved in the water. Conductivity is often measured in units called siemens (S). Its reciprocal, ohm-centimeters (ohm-cm), which measures resistivity, is an optional form of measurement. The lower the conductivity (i.e., the lower the number of siemens), the greater the purity of the water. Because siemens are such relatively large units, the values are more readily expressed in microsiemens (μS). Distilled water has a conductance of 0.5 to 4.0 μS; a conductivity level as high as 4.0 μS is acceptable for most tissue culturing. Levels higher than this are often tolerated in cultures, but the best measure of acceptable water quality is plant performance.

A conductivity meter is an essential tool for any tissue culture operation. A conductivity meter costs about US $400; sometimes a meter is built into the water purification system. Most companies dealing in purified water and purification equipment will test water samples without charge. Some government laboratories will also provide this service.

The 3 most common methods used to remove dissolved chemicals from water are distillation, deionization, and reverse osmosis. Sometimes a combination of 2 or 3 methods is required.

Water purified by *distillation* has been the standard laboratory-grade water for many years. The purity of the end product depends on the feed water and the efficiency of the still. Figure 4-4 shows a Barnstead classic still. Water is boiled in a still, leaving nonvolatile salts in the boiler. The steam passes through a condensing coil, and the condensed distilled water is collected in a container, ready to use. One Barnstead still (product number A440266, MP1), which costs about US $1200, has been referred to as the "tissue culture still." It produces 1.4 liters (about 0.4 gallons) of distilled water per hour. A less-expensive (US $700) compact glass still capable of similar production is available from Carolina Biological Supply Company (70 1699). Sometimes it is necessary to run water through a glass still twice to achieve the desired purity.

Deionization is another effective way to remove dissolved chemicals from water. A single mixed resin bed tank, or cartridge, can provide adequate purity for normal tissue culture production. The system is connected to existing water lines and provides a continuous flow of deionized water. Water passes through a mixture of positive and negative ion-exchange resins. The undesirable ions in the water are exchanged for H^+ and OH^- ions on the resins. These H^+ and OH^- ions combine to form water (H_2O), while the undesirable ions are left on the resins. If the feed water is relatively free of impurities, the resin beds (or cartridges) should work efficiently for several months. Resin bed exchangers are usually equipped with a warning light to indicate when the resins are spent. The light is illuminated when the conductivity level of the product water is becoming too high. If the conductivity of the product water is too high that means that the resins are saturated with undesirable ions. Companies that sell pure water and water purification equipment rent and sell mixed resin tanks, which they will regenerate for US $100 or more, depending on the size. They also sell disposable cartridges.

Various deionizing systems that use disposable cartridges (Figure 4-5) are available from scientific supply companies. One such system (Barnstead's Bantam D0800B) offers a demineralizer, which includes a stand to hold the cartridges, a conductivity meter with indicator, and a connecting hose, all for about US $500. Disposable cartridges to fit this unit include a mixed bed for about US $70 (Barnstead D0809). This unit can remove a total of 104 grams (g) of calcium carbonate. For example, if the feed water contains 81 milligrams (mg) of calcium carbonate per liter, which gives it a resistivity of 5000 ohm-cm (or a conductivity of 200 μS), this unit could purify 1284 liters of feed water. If 10 liters of distilled water are required per day, the cartridge will have to be replaced in 5 months. The direct cost of 1 liter of purified water in this case is about US $0.035, exclusive of the original investment and costs for cartridge renewal and power to operate the unit.

Reverse osmosis, a third method of water purification, is usually used in combination with a deionizer or still. It does not have the refinement of the other 2 methods, but it provides excellent pretreatment. *Osmosis* is the diffusion of a fluid through a membrane into a solution of higher ion concentration to equalize concentrations on both sides of the membrane. *Reverse osmosis* forces the solution through a membrane in the opposite direction, filtering out the impu-

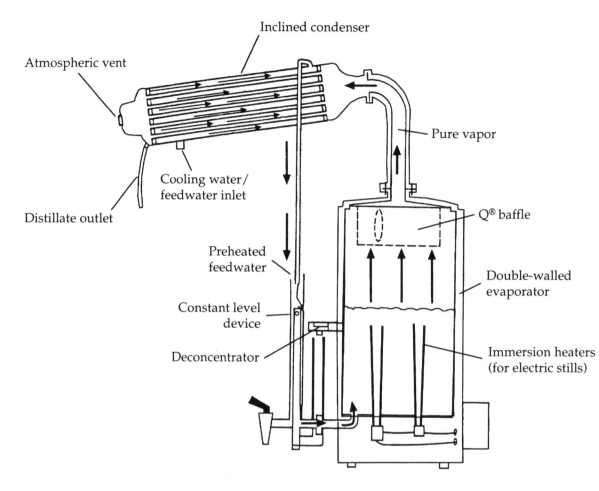

Figure 4-4. Barnstead classic still. From *The Water Book* 1994, reprinted with permission from Barnstead/Thermolyne Corporation, Dubuque, Iowa.

rities. Raw feed water is run through the reverse osmosis system to remove most of the impurities prior to flowing through a deionizer or entering a still. The efficiency of the subsequent still or deionizer is increased significantly when the water first goes through reverse osmosis. Reverse osmosis systems are initially expensive, costing over US $1000, but combined with a still or deionizer, and given the right feed water (nonacidic, low in calcium and iron), extra-high-grade water (0.06 µS) can be obtained. Manufacturers claim that the cost of replacing deionizing cartridges is reduced by a factor of 8 when using reverse osmosis pretreatment.

pH meter

The acidity or alkalinity (pH) of media is crucial in tissue culture and is specific to the requirements of specific plants, just as it is in soils and potting mixes. For example, rhododendrons are tissue cultured in a relatively acidic medium (pH 4.5–5.0), whereas strawberries *(Fragaria)* require a less-acidic medium (pH 5.7). Figure 5-2 shows the range of pH levels.

A commercial laboratory should invest in a pH meter. A pH meter costs from US $260 and up; a high-quality meter will cost in the neighborhood of US $750. A beginner, however, can get by simply with pH indicator paper. Recommended pH ranges for test papers are 2.9–5.2, 4.9–

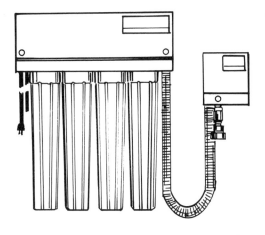

Figure 4-5. A series of water purification cartridges.

6.9, and 5.5–8.0. These small ranges will measure adjustments of media to approximate pH requirements.

Balance

Unless premixed chemicals are to be used, a precision balance will be required to accurately measure some of the small amounts of the chemicals required for tissue culture media, although the accuracy required of analytical chemists is not usually required in plant tissue culture. Most analytical balances cost over US $2000. These expensive, electronic one-pan balances are fast and precise and are a worthwhile investment if affordable.

An Ohaus brand portable balance (CT200) with 0.01-gram accuracy is available for about US $600. Though less precise than analytical balances, it is adequate for weighing quantities of 0.01 to 10 g (10 to 10,000 mg).

The dilution method is one alternative for weighing smaller amounts. By this method, if, for example, 10 mg of a compound are required per liter of medium, place 100 mg (0.1 g) of the compound in 100 milliliters (ml) of water. Stir it well to be sure the compound is properly dissolved in the solution, then use 10 ml of this solution in your liter of medium. The remaining solution can be refrigerated for future use.

When weighing amounts of more than 10 g, use a triple-beam balance, saving the more sensitive balance for the smaller quantities. A good triple-beam balance costs about US $150 (Ohaus Dial-O-Gram).

Hot plate/stirrer

Agar, a gelatinous polysaccharide extracted from certain red algae, is required as a solidifying agent in many media formulas. In media preparation, agar in powder form is added to the water together with the other chemicals. This mixture must be boiled briefly to melt the agar and dissolve the chemicals to form a homogeneous mixture. Unless the mixture is constantly and effectively stirred until boiling—agar melts at boiling temperature (212°F [100°C])—the agar will settle and stick to the bottom of a flask and burn. For that reason, a combination hot plate and automatic stirrer is among the most useful tools for preparing media. A rotating magnet built into the hot plate/stirrer causes a magnetic stir bar, placed inside the flask, to rotate and so prevent sticking (Figure 4-6). The stirrer and hot plate features can be used at the same time or independently. The stir bar should be used in heat-resistant glass containers on the hot plate; metal containers generally will not work on a hot plate/stirrer because the stir bar is magnetic. An additional and very important use for the automatic stirrer (without the hot plate feature) is to provide agitation during the cleaning of explant material. A small hot plate/stirrer costs about US $260.

Figure 4-6. A hot plate/stirrer, showing the rotating magnetic bar in a flask of medium.

A gas burner or an electric stove top will heat media much faster than a hot plate, so it is more efficient for larger quantities of media. If you use a gas burner you will need a large non-aluminum cooking vessel, possibly a canner, and a means of stirring. An electric kitchen beater is an inexpensive option. It can be mounted on a board so that it does not have to be hand held (Figure 4-7). A preferred but more costly option is to buy a top-mounted motor and stirring rod, which costs about US $300.

A hand-held stirring rod, beater, or spoon will also work for stirring media, but there is a greater likelihood of the agar sticking, unless the solution is in a double boiler (the pan of agar set within a pan of water). There is also a greater hazard of being burned should a hot medium splatter.

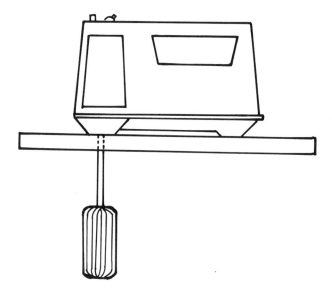

Figure 4-7. A hand mixer (beater) mounted on a board for mixing media in a pan on a stove.

Another option for melting agar is especially useful for larger amounts of medium. For 3 or more liters of medium, place a flask containing agar and 1 liter of water in a microwave oven, stirring occasionally. When the agar is melted, add the solution to the medium, which has already been mixed and heated except for the agar. Be sure to allow room for the agar solution in the 3 liters of medium.

Media dispenser

A small laboratory can easily forego high-priced equipment for dispensing media. A 10-ml polypropylene pipet can be purchased for about US $16. A simple heat-resistant Pyrex pitcher will serve very well to pour a few liters of a hot medium into test tubes or jars, and it costs only US $6. A coffee urn can also sometimes be used. Or use a reservoir for media that is gravity fed through tubing, with a pinch clamp to control the flow.

An automatic pipetter to dispense media is a labor-saving device for large-scale production. Whereas a good technician with a quick eye and a steady hand can pour 100 test tubes in 10 minutes from a 1-liter pitcher, the automatic dispenser will fill 500 tubes in the same amount of time. An automatic pipetter costs about US $1100.

Explant cleaning equipment

A hot plate/stirrer (with the heat off) usually provides sufficient agitation for cleaning explants; however, a tightly closed jar containing explants and cleaning solution, shaken by hand, is a very effective method as well. Alternatively, a mechanical shaker or rotator (see below) can be built or purchased and will provide effective agitation over long periods of time.

A few laboratories use vacuum pumps to help disinfect explants. The suction of the vacuum (25 millimeters of mercury) improves contact between the cleaning solution and the explant and helps to disengage contaminants. A vacuum pump or aspirator is also useful for ultra-filtration of heat-labile chemicals. The need for a vacuum pump, or aspirator pump, should be established before investing in one because they are cumbersome to use, and too high a vacuum can injure explants.

An ultrasonic cleaner is another option to help remove contaminants. The action of the ultrasonic cleaner causes different materials to vibrate at different rates, thus dissociating them. The contaminants are literally shaken loose from the plant material. A small ultrasonic cleaner costs about US $100. Here again, caution is advised or the tissues will be injured.

Sterilizing equipment for media

A modest laboratory with no more than 3 people can readily use a household pressure cooker (Figure 4-8) for sterilizing test tubes or jars containing media. A 21-liter (5.5-gallon) pressure cooker costs less than US $100. A burner or stove-top unit is necessary for heating the cooker. A wire basket (Figure 4-8) to hold the test tubes must be built or purchased when using such a cooker. The basket consists of a wire frame with handles, lined with 0.5-in (or 1-cm) hardware cloth (wire mesh) to prevent the test tubes from falling out.

Autoclaves, a common method of sterilization for tissue culture, come in many sizes. Most are automated, more efficient, and more accurate than cookers. A larger operation should have an autoclave, but unless the size and automation of the autoclave appear to provide significant savings in labor, you may as well use pressure cookers. Autoclaves cost from about US $5000 for a chamber 10 × 18 in (25 × 45 cm), to US $49,000 for one with a chamber 20 × 38 in (50 × 95 cm). The size of the autoclave needed can be calculated by estimating the number of containers to be processed in a day, how many containers will fit in a particular autoclave, and the length of the total processing time (heating, holding, and shutting down) for that autoclave.

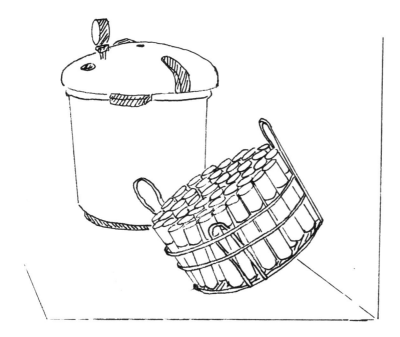

Figure 4-8. A pressure cooker or canner (*background*) and a wire basket containing test tubes (*foreground*). The basket and test tubes are slanted after sterilization so that the surface of the medium will be slanted when it cools and solidifies.

Some organic chemicals, however, are heat labile (unstable) and will break down when exposed to the heat of certain sterilization procedures. Therefore, some media ingredients must be sterilized by cold sterilization using a special syringe and filter. The problem of heat lability varies with the chemical, the temperature, and the duration for which the temperature is applied. Cold sterilization is accomplished in the sterile environment of the transfer hood. A solution of the chemical is loaded into a syringe, and a sterile membrane filter unit is then attached to the syringe. The solution is forced through the membrane filter into a sterile container, ready to be added to a sterilized medium, whether liquid or agar, which has been cooled but not solidified so that the additive can still be mixed in.

Some people cold sterilize hormones and vitamins before adding them to sterilized media, but this is usually not necessary. The major application of cold sterilization is in the use of antibiotics, most of which are very heat labile. Whatever sterilization method is used, whether hot or cold, it is important to not over-sterilize the medium.

Sterilizing equipment for transfer tools

In the past, tools used in the transfer hood primarily were sterilized by dipping them in alcohol, followed by flaming (burning off the alcohol) using a Bunsen burner or alcohol lamp. Alcohol used for disinfecting instruments in this manner should be 95% ethyl or isopropyl alcohol, because at 95% strength it will burn off more easily than will a more dilute form. Do not use methyl alcohol because it is extremely poisonous, even just to breathe the fumes.

The Touch-O-Matic is a modern modification of the old familiar Bunsen burner and costs about US $105. It saves gas and minimizes the danger of an open flame. A pinpoint pilot flame burns continuously and the higher flame is lit by resting one's hand on a disc while flaming an instrument. A primitive alcohol lamp can also be improvised by using a covered container and a wick, or it can be purchased for about US $9. These burners all involve open flames, which is hazardous especially where alcohol is used for disinfection. They are less safe, but also less expensive, than some of the more modern sterilizing equipment.

For sterilizing implements in the transfer hood, the preference today is the glass bead sterilizer, an insulated pot with a component that heats glass beads within it (Figure 4-9). A glass bead sterilizer costs about US $415 (cheaper ones for about US $250 can be purchased from India). An inexpensive alternative to the glass-bead principle is a lead melting pot, which is filled with sand. It sells for about US $45 (Midway 637732). The lead melting pots are not well insulated and can burn out in less than a year.

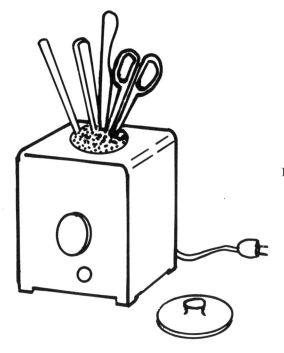

Figure 4-9. A glass bead sterilizer.

Another choice used by a large number of laboratories today is the Bacti-Cinerator (Figure 4-10), an infrared sterilizer that costs about US $250 (Fungi Perfecti B-111). Replacement heater elements, which may be needed once or twice a year, cost about US $70. The Bacti-Cinerator sterilizes instruments in 5 seconds at 1600°F (870°C) as the instruments are inserted into the red-hot, hollow cone of the cylinder. The glass bead sterilizers are similarly heated.

Figure 4-10. A Bacti-Cinerator sterilizer.

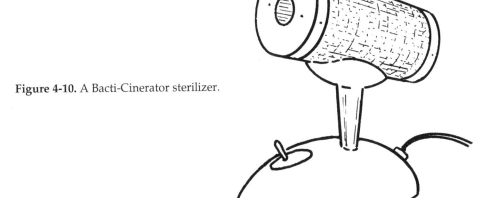

Bleach solutions are satisfactory for sterilizing tools in some situations. Two concentrations of bleach solutions should be used with this method: a 10% solution (1 part bleach [5% sodium hypochlorite] in 10 parts solution, as in 10 ml of bleach mixed with 90 ml of water; also referred to as 1/10 bleach) to soak the instruments and a 1% solution (1 part bleach in 100 parts solution) to rinse them. This is the method used in the procedures described in this book because it is the least expensive option and serves as a good teaching tool. Remember, however, that bleach will burn the skin and the fumes are undesirable.

Rotator/Shaker

Some cultures grow better in liquid media than on solid agar media. Such cultures often need to be gently agitated, which aerates the medium and thereby prevents the culture from "drowning." Agitation also disperses the waste products of the culture. Cultures can be gently agitated by means of a rotator (Figure 4-11) at 1 rotation per minute (rpm), or by a shaker. Shakers are motorized to provide a back-and-forth motion.

Agitation is particularly useful if you are attempting to grow callus tissue. The mixing disorients the tissue and, in so doing, inhibits plantlet formation while allowing for increased callus production.

One convenient orbital shaker (Cole-Palmer G-04726-20), which lacks a special test tube rack but has a 10 × 12 in (25 × 30 cm) platform, variable speed, timer, and audible alarm, costs US $425. Others range from less than US $200 to over US $2000, depending on the size and whether or not racks for holding test tubes or flasks are purchased as well.

A rotator or shaker is also useful for cleaning explants, especially for any overnight treatment. A rotator to hold about 75 test tubes can be built for less than US $150 and a little mechanical ingenuity.

Figure 4-11. CEL-GRO Tissue Culture Rotator (Model No. 1640). Courtesy LAB-LINE® Instruments, Inc., Melrose Park, Illinois.

Dishwasher

A conventional built-in household dishwasher, costing about US $500, will handle the dishwashing requirements of a small laboratory (fewer than 2000 test tubes per day). If the prongs on the bottom rack of the dishwasher are bent down, 4 test tube racks each holding 40 test tubes—160 test tubes total—can be washed at one time. Tubes in racks must be covered with wire mesh and turned upside-down in the dishwasher. A commercial laboratory dishwasher of the same capacity costs at least 4 times as much as a kitchen dishwasher and yields few advantages. However, a large commercial washer should be considered when the volume of glassware used warrants such a purchase.

Refrigerator

A household refrigerator is useful to store perishable chemicals and stock solutions. A larger refrigerator is necessary if media or cultures are to be refrigerated as well. (No food or beverages are allowed in laboratory refrigerators by government mandate.)

A walk-in refrigerator is a valuable means of treating or delaying the growth of cold-resistant cultures. Chilling is often called for if plantlets are ready too soon to be transferred to the greenhouse, or if stock cultures need to be slowed down. Cultures of some hardwood species benefit from cold treatment before acclimatization.

An expansion coil with fan, compressor, and temperature controller are basic off-the-shelf components that can be sized to your needs. A 24×24 in (60×60 cm) exchanger (expansion coil) and a ¾-horsepower compressor will refrigerate an insulated room $10 \times 10 \times 7$ ft ($3 \times 3 \times 2.1$ m).

Labeler

Different varieties of plants are not easily identifiable in culture, and it is much more difficult to distinguish one culture from another than it is to tell one plant from another in the greenhouse. It is extremely important, therefore, to identify each container at the time it is processed. A practical labeling procedure is essential, but most systems do have their drawbacks. Although marking pencils will work, it is a slow process and any moisture on the glass or the pencil makes writing difficult. If using a marking pen with permanent waterproof ink, the ink must be scrubbed off with a pot scrubber before the container can be reused. If using a marking pen with ink that washes off (as in an overhead-projector marker), the ink may rub off prematurely. Hand stamps are also a problem due to permanent ink. Both markers and hand stamps are too slow for large-scale commercial production. Grocery-store price labelers prove satisfactory in most cases. The least expensive kinds usually cost between US $100 and $200. If labels come off in the dishwasher, however, the dishwasher may clog. There is no one solution to fit every situation.

Microscope

The decision to buy a microscope depends on the type of tissue culture work to be done, as well as on the interest and curiosity of the technician. A stereo dissecting microscope is usually required for excising meristems. A microscope with 20× or 40× magnification costs about US $400. Cheaper magnifiers are available and may meet your particular needs. Some people with exceptionally good eyesight may not need a magnifier at all for excising meristems. A compound microscope is useless for excising meristems because there is not enough working room between the objective and the specimen.

A compound microscope is not necessary for routine commercial production, but one is essential if you plan to identify contaminants (see Chapter 11). Student monocular compound microscopes are available for about US $500. Keep in mind that the cheaper the microscope, the more likely it will have poorer resolution or sharpness. *Resolution* can be defined as the ability of a microscope lens system to give individuality to 2 objects (e.g., 2 bacterial cells) that lie in close proximity to one another. For more information on microscopes, see Appendix B.

Transfer chamber or hood

Laminar air flow transfer hoods (Figures 4-12 and 4-13) are essential for commercial operations. They provide a sterile atmosphere in which to work with cultures. Air is forced through a HEPA filter, located at the back of the hood, that strains out particles as small as 0.3 micrometers (μm). The gentle, scarcely detectable air stream flows through the filter, across the work area toward the technician (at 100 ft [30 m] per minute), providing a sterile atmosphere in which the technician works.

A popular bench-top laminar air flow hood to accommodate one technician can be purchased for US $950 (Fungi Perfecti LFH-II). Fungi Perfecti also sells a kit, consisting of all the necessary components with which to assemble a hood, for US $425, and one that is wide enough for 2 technicians is available for US $1200. Other laboratory supply companies sell assorted designs of laminar flow hoods, costing up to US $4000 or more. Some larger units that set on the floor have the option of air intake either at floor level or at the top of the unit. The unit that has air intake on top rather than at floor level is preferred because the air at floor level is likely to have more contaminants than the air nearer the ceiling.

HEPA filters can be purchased separately from the hood. A 24 × 36 in (60 × 90 cm) HEPA filter costs less than US $300; a 12 × 12 in (30 × 30 cm) filter costs about US $90 (Fungi Perfecti MR 1212).

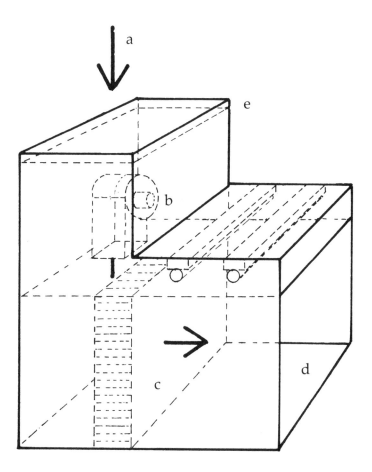

Figure 4-12. A laminar air flow transfer chamber: a. air flow; b. blower; c. HEPA filter; d. work area; e. replaceable prefilter (furnace filter).

Figure 4-13. Detail of work area of a laminar air flow transfer chamber.

It is possible to construct, rather than buy, a satisfactory laminar flow hood, or have a local cabinet shop make one. Additional components necessary for building a transfer hood include prefilters (inexpensive furnace filters), a blower, plywood or plexiglass sides and top, a smooth bench top, and a fluorescent light.

A still-air transfer chamber with a slanted glass front and partially enclosed hand access is an alternative for the hobbyist, especially if home built (see Figure 4-1). Some hobbyists have worked successfully with large clear plastic bags. Specially designed, commercially available bags, into which sterile air can be blown, come with built-in sleeves. AtmosBags from Sigma Chemical cost about US $28 for a medium size, which is $24 \times 39 \times 48$ in ($60 \times 98 \times 120$ cm) before inflating, or about 17,000 cubic inches (280 liters) when inflated.

A kitchen oven has also been used successfully as an alternative. The oven is first heated for sterilization, allowed to cool without opening the door, and then protected by a plastic curtain in front. Reaching in with gloved hands may be awkward, depending on the height of the oven.

Growing room shelves

Lighted shelves need to be constructed in the growing room to hold the test tubes or jars of cultures while they are growing. Shelves can be built using slotted steel angle supports for 4×8 ft (1.2×2.4 m) shelving of particle board, plywood, or wire mesh of expanded metal, or 0.25- to 0.5-in (about 0.5- to 1-cm) hardware cloth. Wire mesh is more expensive than solid shelving but allows better air circulation. Boards should be painted white to maximize available lighting. A room 8 ft (2.4 m) high will accommodate 5 shelves placed 18 in (45 cm) apart, with the bottom shelf 4 in (10 cm) from the floor (this distance underneath provides room for cleaning). One rack, consisting of particle-board shelves, metal framing, and lights, can be purchased for about US $700.

Additional supplies

Household items often can substitute for traditional glassware and tools used in the laboratory (Figure 4-14). Such items are appropriate if they are heat and pressure resistant, to allow for sterilization. Heat-resistant Pyrex beakers are good vessels for cleaning explants. Aluminum utensils are undesirable because they can give off aluminum ions into the medium.

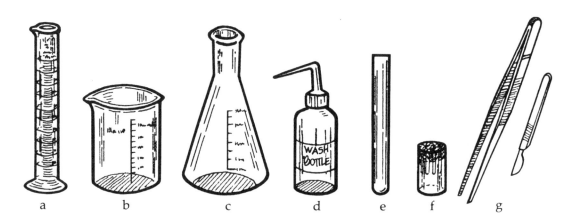

Figure 4-14. Typical laboratory glassware and tools: a. graduated cylinder; b. beaker; c. Erlenmeyer flask; d. wash bottle; e. test tube; f. closure (cap) for test tube, showing cotton filter; g. forceps and knife (scalpel).

Pint-size Mason canning jars are readily available containers in which to grow plant tissue cultures (Figure 4-15). They are inexpensive, convenient, autoclavable, and easy to handle. The typical metal lids do not allow sufficient light for the cultures, so they must be substituted with some type of transparent lid. Rounds of plate glass are generally too expensive and cumbersome for this purpose. Polycarbonate, an autoclavable plastic available in 20-mil (0.05-mm) sheets, is suitable for cutting rounds to fit the tops of pint jars. For media preparation, the metal rings that come with Mason jars can be used to hold the rounds in place. When transferring cultures, remove the ring and the polycarbonate round, placing the polycarbonate lid on a sterile paper towel. After positioning the transfers in the jar, replace the polycarbonate lid and place a square of plastic wrap or a small sandwich bag over the top, and then screw on the metal ring over the wrap. Plastic closures for pint-size Mason jars, made especially for tissue culture purposes, are also available commercially, but they are expensive (more than US $1) and not entirely contaminant proof; they can be a good investment, however, depending on sanitation levels and frequency of transfer.

Baby-food jars provide another convenient and inexpensive vessel for tissue cultures. They too have commercially produced closures available. They can be used instead of, or in addition to, pint jars or Magenta GA-7's (autoclavable polycarbonate containers for tissue culture).

Test tube closures do not offer maximum protection against contaminants. They should be lined on the top end with a 0.25-in (0.5-cm) thick piece of nonabsorbent cotton (illustrated in Figure 4-14f). Do not use absorbent cotton because it will take up moisture, thus making it more conducive to contamination.

An instrument holder is useful in the transfer hood, especially when using a Bacti-Cinerator sterilizer. An effective instrument holder can be made using a $6 \times 22 \times 0.5$ in ($15 \times 55 \times 1$ cm)

Figure 4-15. Mason pint jar, with a polycarbonate disc under plastic sandwich wrap and secured with a rubber band.

piece of 0.75-in (1.5-cm) hardware cloth. First form the hardware cloth so that the back is 10.5 in (26 cm) high, the front 9.75 in (24 cm) high, and the base between is 2.25 in (5.5 cm) wide. Place in a cardboard or wooden mold 2 in (5 cm) deep and 2.25 in (5.5 cm) wide, and pour in wet plaster of Paris (hemihydrate of calcium sulfate). When dry, remove from the mold.

A metal test tube holder standing on end makes a more immediate but also more expensive holder. Do not buy a plastic-coated holder because it will melt with the heat from the instruments. A zinc-plated, steel wire rack measuring about $10.5 \times 4.5 \times 3$ in ($26 \times 11 \times 7.5$ cm) costs about US $20 (VWR Scientific 60940-026); an example of such a metal rack is shown in Figure 4-16a. This is less stable than the homemade holder described above, however. Another option is to make an instrument holder out of glass tubing by heating it and bending it to shape.

Seedling containers (Speedling trays, Todd planters, or others; Figure 4-16c) are useful for holding test tubes in the growing room because they allow the test tubes to be slanted, which permits exposure to more light than if the test tubes are upright, as they are in the typical wire-rack test tube holder. The 40-tube holders (Figure 4-16a) are almost essential for convenience in washing, storing, and other handling of test tubes. A small holder for 10 test tubes (Figure 4-16b) is convenient in the transfer chamber or when handling small numbers of tubes. This type of holder can be built with plywood.

USE AND CARE OF EQUIPMENT

Remember that laboratory precision instruments are delicate and demand very careful use—not like tractors or refrigerators, which can stand a certain amount of abuse. Consequently, it is of utmost importance that you read and carefully follow the instructions provided by the equipment manufacturers before you touch a new instrument. If questions arise, direct them to the vendor or the manufacturer. It is in their best interest to have informed, satisfied customers; and it is in your best interest and to your advantage to know how to operate and care for your investment.

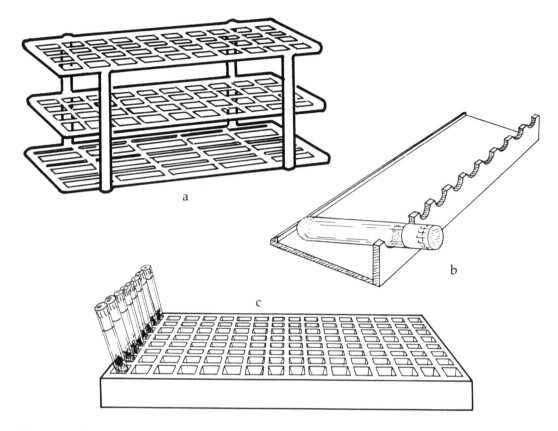

Figure 4-16. Common varieties of test tube holders: a. wire-rack test tube holder for 40 tubes; b. plywood test tube holder for 10 tubes; c. Speedling or Todd planter tray used as test tube holder for 128 tubes.

SUPPLY LIST

Most of the suggested supplies listed in Table 4-1 (pages 58–59) are for a modest 3-person operation (about 200,000 plants per year). Your individual selections will depend on your budget and your goals—some supplies you may choose to forego altogether and for others you may prefer to use homemade alternatives. Some of the items listed are available from supermarkets. Most of the items can be purchased from scientific supply companies (see Appendix D); a business license may be required to purchase from some suppliers. The prices listed are in 1995 US$ and are what you might expect to pay for an average-quality product; prices may vary.

Table 4-1. Equipment and Supplies

Item	Size	Number	US$, Total	Use
Alcohol, isopropyl or ethyl, 70%	1 pt	2	1.50	Clean explants, work surfaces
Aluminum foil, heavy	25-ft roll	1	2.50	Cover items for sterilizing
Autoclave	10 × 18 in	1	5000.00	Sterilize media
Balance, electronic (Ohaus CT200)	n/a	1	600.00	Weigh small amounts
Balance, triple beam	n/a	1	150.00	Weigh amounts over 10 g
Beaker (Pyrex)	150 ml	12	28.00	Clean explants
Beaker	250 ml	12	28.00	Mix chemicals
Beaker	600 ml	6	21.50	Miscellaneous dispensing
Bleach (5% NaOCl)	1 gal	1	1.00	Clean explants, instruments, etc.
Bottle, spray	8 oz	2	5.00	Mist for Stage IV culture
Bottle, wash	125 ml	6	12.50	Rinse pH probe
Brush, flask	16 in	1	9.00	Clean flasks
Brush, test tube	10 in	10	20.00	Clean test tubes
Conductivity meter	n/a	1	400.00	Check ion content of water
Cotton, nonabsorbent	1 lb	1	14.50	Line top of lids
Cylinder, graduated	10 ml	1	7.00	Measure liquids
Cylinder, graduated	100 ml	1	9.00	Measure liquids
Detergent, liquid	1 qt	1	1.50	Miscellaneous cleaning
Detergent, powder	1 lb	4	3.00	Dishwasher
Dish, petri, Nalgene	100 mm	10	35.00	Microdissection; research
Dish, plastic, glass, or ceramic	2 × 6 × 9 in	2	6.00	Hold bleach and tools in hood
Dish, sterile, disposable	100 mm	500	65.00	Microdissection; callus, fungus growth
Dishwasher	n/a	1	500.00	Wash glassware
Dropper, medicine	1 oz	12	3.25	Dispense small liquid amounts; adjust pH
Filter, furnace	18 × 25 × 1 in	6	6.00	Prefilters for transfer hood
Flask, Erlenmeyer	1000 ml	6	36.65	Mix, store stock solutions
Flask, Erlenmeyer	2000 ml	1	15.60	Mix media
Flask, Erlenmeyer	4000 ml	1	39.90	Mix media
Forceps, stainless	10 in	2	28.00	Transfer cultures
Forceps, stainless steel (tweezers)	4.75 in	1	5.25	Microdissection
Gloves, surgical	any	100	12.00	Protect hands
Hot plate/stirrer	n/a	1	260.00	Heat, mix media; clean explants
Jar, baby-food	6 oz	100	32.00	Grow cultures
Lids for baby-food jars	n/a	100	30.00	Cover jars
Jar, canning (Mason)	1 pt	120	50.00	Grow cultures
Jar holder	n/a	1	3.00	Move hot jars
Knife, plastic handle	6 in	4	4.00	Divide cultures
Labeler, grocery-store type	n/a	1	100.00	Label cultures
Light meter	n/a	1	90.00	Measure light distribution, intensity
Magenta containers	77 mm	100	170.00	Grow cultures
Microscope, stereo	n/a	1	400.00	Observe meristems for excision
Mop, sponge	n/a	1	11.00	Clean floors
Pens, marking, waterproof	n/a	1	1.00	Permanent label
Pens, marking, not waterproof	n/a	2	3.00	Temporary label
pH meter	n/a	1	500.00	Measure pH
Pipet, volumetric, polypropylene	10 ml	1	16.00	Miscellaneous dispensing
Pipet filler	n/a	2	9.00	Pipet safety
Pitcher, Pyrex	1 qt	1	6.00	Dispense media
Polycarbonate sheets	20 mil, 2 × 4 ft	1	7.80	Lids for jars

Item	Size	Number	US$, Total	Use
Pot holder, kitchen	6 × 6 in	4	8.00	Hold hot items
Pressure cooker	n/a	1	100.00	Sterilize media
Refrigerator	n/a	1	500.00	Store organics, cultures, etc.
Scalpel, stainless	8.5 in	2	74.00	Dissection and transfer
Scoop or spoon	2 oz	2	1.00	Scoop compounds for weighing
Shaker	n/a	1	425.00	Agitate liquid media; clean explants
Spatula, stainless	200 mm	3	11.00	Scoop compounds for weighing
Sponge, household	medium	4	2.00	General cleaning
Sterilizer, Bacti-Cinerator	n/a	1	231.00	Sterilize tools in hood
Sterilizer, glass bead	n/a	1	415.00	Sterilize tools in hood
Sterilizer, Touch-O-Matic	n/a	1	106.65	Sterilize tools in hood
Still, water	n/a	1	500.00	Purify water
Syringe, B-D sterile	10 ml	100	23.00	Cold sterilization
Syringe filtration, 50-unit pkg. (Gelman)	0.45 µm	1	82.00	Cold sterilization
Test tubes (culture tubes)	25 × 150 mm	80	50.00	Grow explant starts
Test tube closures	25 mm	100	14.00	Lids for test tubes
Test tube rack, plywood	holds 10 tubes	2	8.00	Hold tubes in hood
Test tube rack, wire	holds 40 tubes	2	40.00	Hold tubes for washing, storing
Thermometer	−20 to 110°C	2	8.00	General purpose
Timer, household	60 min	2	26.00	Time cookers
Timer/controller	24 hr	1	75.00	Control lights in growing room
Towels, paper	roll	6	3.00	Dry hands, spills
Towels, paper, single-fold	9.5 × 10.25 in	4000	30.00	Cut cultures on
Tray, household	10 × 16 in	8	24.00	Hold jars
Tray, seedling, 128 cell	11 × 21.25 × 2 in	100	70.00	Hold test tubes; plantlets in Stage IV
Tray, Speedling, 128 cell	n/a	6	30.00	Hold test tubes in growing room
Wastebasket, large	n/a	1	6.00	Media room
Wastebasket, medium	n/a	1	4.00	Transfer room
Weighing papers	3 × 3 in	1 pkg	10.00	Hold chemicals for weighing

Chapter 5: Media Ingredients

The standard formulas for tissue culture media have been determined by research scientists to provide optimum nutrients and growth regulators for specific plants. The formulas developed by Toshio Murashige and his associates, particularly MS (Murashige and Skoog's) medium, are probably the best-known standard formulas, and which are used primarily for herbaceous foliage plants. The woody plant medium (WPM), developed by Brent McCown and Greg Lloyd, is designed, as the name implies, to optimize tissue culture of certain woody plants.

Combining the various chemicals for the media in which plant cultures grow is an art, similar to that of cooking a meal—the ingredients and language are different, and some of the equipment may not be familiar, but recipes (formulas) usually need to be followed in a precise, orderly manner. The nutrients in media affect the health, vigor, and growth of the specialized cells, tissues, and organs as the plantlets differentiate and develop in culture.

As an alternative to mixing the numerous chemicals yourself, ready-mixed powdered media, available in various stages of completeness, can be purchased from several supply companies (see Appendix D). Using a premixed medium is comparable to making a cake from a packaged mix as opposed to making one from scratch. Even though many premixed media are available, most people choose to do their own mixing of formulas because it is less expensive and it is easier to adjust the formula in response to culture performance.

Perhaps you were bored with chemistry in school. Here in the field of micropropagation, however, the elements take on a whole new dimension as they relate and translate directly to plant response—a dramatic and exciting process. Read or skim this chapter for what it can mean to you. If a cupboard full of unfamiliar ingredients is too overwhelming, then buy premixed media, at least to begin with. If the chemistry review is meaningful to you, accept the challenge and put your curiosity to work because the whole world of micropropagation is open to you. As a gardener or plant propagator you will be in somewhat familiar country because you have grown plants, used fertilizers, and probably used hormones to root cuttings. With a can-do attitude, you will learn to enjoy the successes and steer clear of the failures.

Without at least a beginning course in chemistry, it may take you a while to become familiarized with the chemical language, symbols, and working principles of mixing media. Initially, the number of chemicals may seem excessive. If you are a novice, however, by starting with a commonplace plant for which a medium formula has been well researched and applied, the task becomes relatively simple. A cookbook approach fortified with determination will help the unsure beginner gain confidence.

Whether mixing the media for yourself or using premixed media, it is useful to know something about chemicals and their role in plant growth and development. The information presented here should help the beginner appreciate the art and science of media formulation. The expert may choose to scan this section for review. Many of the terms and concepts used by chemists are household words; however, a brief review of definitions is desirable to strengthen the foundation for formula mixing.

CHEMISTRY REVIEW

The metric system

The metric system is the system of measurements used in science, and consequently, it is used in tissue culture. Measurement calculations are simpler than in the English system because the metric system is based on multiples of 10. Committing to memory certain relationships and terms will make the metric system easier for the individual trained in the English system. Here are a few crucial ones: there are 1000 milligrams (mg) in 1 gram (g), 1000 milliliters (ml) in 1 liter, 10 millimeters (mm) in 1 centimeter (cm), and 1000 micrometers (μm; formerly called microns, μ) in 1 mm. Another important equivalency is 1 cubic centimeter (cm^3) of water has a volume of 1 milliliter and weighs 1 gram. For comparison with some English system measurements, 1 gram is equivalent to about 0.035 ounces (oz), 1 liter is about 1.1 quart (qt), and 1 centimeter is about 0.4 inches (in). For a further discussion of the metric system, see Appendix A.

Weights and measures

For weighing chemicals it is useful to have an understanding of the concept of significant figures. Simply stated, when weighing chemicals the larger the total amount required, the less important precision and accuracy become. One milligram is an insignificant amount when considering *100 grams;* but when *10 milligrams* are required, 1 milligram becomes a significant factor, amounting to 10% of the whole. This concept of degree of importance is referred to as significant figures, and it applies to both weights and volumes. Three significant figures in the final measure is adequate for production work; in other words, if, for example, a formula calls for 184.4 g of a compound, 184 g (3 figures) is sufficiently accurate, and 15.6 mg is satisfactory if 15.57 mg are called for. This point must be made because often more than 3 figures are given in the literature. People involved in research need to be extremely precise, but people in production usually need not be quite so accurate. Remember this when you are adjusting formulas and the calculator spins out 8 decimals (Figure 5-1).

Some laboratory glassware, particularly beakers and Erlenmeyer flasks, carry the caution that the calibrations on the side are accurate only to within plus or minus 5%. Accuracy within 5% is usually sufficient for our purposes.

Basic units of chemistry

An *atom* is the smallest particle of an element that retains the chemical characteristics of that element. The number of atoms in a chemical formula is always written as a subscript, below and to the right of the letter symbol, as in H_2 or O_2.

An *element* is a substance composed of atoms of the same type and which cannot be separated (decomposed) into simpler substances by any usual chemical means.

A *compound* consists of 2 or more different elements chemically combined in fixed proportions. Some examples of compounds include: water (H_2O), composed of hydrogen and oxygen; hydrochloric acid (HCl), which is hydrogen and chlorine; and sucrose ($C_{12}H_{22}O_{11}$), composed of carbon, hydrogen, and oxygen.

A *molecule* is the smallest quantity into which a chemical compound can be divided and still keep its characteristic properties. The simplest molecules have only one atom. One molecule of water (H_2O) has 3 atoms, 2 hydrogen and 1 oxygen.

An *ion* is an electrically charged atom or group of atoms. For example, table salt, which is sodium chloride (NaCl), *ionizes* (chemically separates) in water to form positively charged sodium ions (Na^+) and negatively charged chloride ions (Cl^-). When water containing calcium

Figure 5-1. Measurements to 3 significant figures are adequate for production work. The measurement here could be rounded off to 9.88.

carbonate ($CaCO_3$) is deionized for purification (see Chapter 4), the positive calcium ions (Ca^{2+}) are removed by a negatively charged resin bed and the negative carbonate ions (CO^{2-}) by a positively charged resin bed.

Atomic and molecular weights

The occasional need to use atomic and molecular weights in tissue culture warrants an introduction to the topic. Atoms of different elements have different weights. *Atomic weight* (aw) is the relative atomic weight of an atom compared to that of oxygen (aw 16; see Table 5-1). The sum of the atomic weights in a molecule is the molecular weight (MW) of a compound.

Sometimes media formulas found in the literature give quantities in molar units rather than in grams per liter, and so an understanding of atomic and molecular weights will help you to find the grams per liter equivalent and ensure correct measurements. A one *molar* (1 *M*) solution is 1 gram molecular weight (the molecular weight of a substance expressed in grams; also called a *mole*) in 1 liter of water. Take, for example, a 1 *M* solution of sodium hydroxide (NaOH), a compound often used for raising the pH of media. Using Table 5-1, we see that the atomic weight of sodium (Na) is 23, that of oxygen (O) is 16, and that of hydrogen (H) is 1; the molecular weight of NaOH is 23 + 16 + 1 = 40. Therefore, 40 g of NaOH in 1 liter of water yields a 1 *M* solution of NaOH. (To be precise, the water plus the NaOH must equal a liter.) Some organic chemicals are very complex, and it is difficult to determine their molecular weight. Fortunately, molecular weights of compounds can be found in catalogs of some chemical companies (such as Sigma Chemical), or see Donnelly and Vidaver (1988).

Another instance in which an understanding of atomic weights is useful is with respect to the water of hydration. Water of hydration is the variable amount of water that is chemically attached to some compounds. The compound manganese sulfate ($MnSO_4$, MW = 151), for example, is manufactured with either 4 molecules of water ($MnSO_4 \cdot 4H_2O$, MW = 223) or as the monohydrate ($MnSO_4 \cdot H_2O$, MW = 169). If a medium formula calls for 2.23 grams of $MnSO_4 \cdot 4H_2O$, then 1.69 grams of $MnSO_4 \cdot H_2O$ would provide an equivalent amount of $MnSO_4$ when using the monohydrate form of the compound ($169 \div 223 \times 2.23 = 1.69$). Such ratios can be used to calculate other equivalents.

**Table 5-1. Atomic Weights of Elements
Commonly Used in Tissue Culture**

Element	Symbol	Atomic Weight
Boron	B	10.811
Calcium	Ca	40.08
Carbon	C	12.01115
Chlorine	Cl	35.453
Cobalt	Co	58.9332
Copper	Cu	63.54
Hydrogen	H	1.00797
Iodine	I	126.9044
Iron	Fe	55.847
Magnesium	Mg	24.312
Manganese	Mn	54.938
Molybdenum	Mo	95.94
Nitrogen	N	14.0067
Oxygen	O	15.9994
Phosphorus	P	30.9738
Potassium	K	39.102
Sodium	Na	22.9898
Sulfur	S	32.064
Zinc	Zn	65.37

Media formulas published in technical journals are sometimes specified in millimoles (mM; 0.001 gram molecular weight) per liter or micromoles (µM) per liter instead of grams per liter. In this case, it is necessary to find the molecular weight of the compound and multiply the moles required by the molecular weight. For example, if 3 mM of calcium chloride ($CaCl_2$) are required per liter of medium, multiplying the molecular weight of $CaCl_2$ (110) by the number of millimoles required will give the required amount of $CaCl_2$ per liter in milligram (mg) units: 110×3 mM = 330 mg. The advantage of expressing media ingredients in moles is that molecules and ions interact as entities; thus, comparing moles instead of weights provides a more valid basis of comparison of one formula with another.

Principles of acids and bases—solutions for pH adjustment

The symbol pH designates the degree of acidity or alkalinity of a solution as indicated by the concentration of hydrogen ions; the higher the hydrogen ion concentration, the greater the acidity. Different plant species require different pH levels in media for optimum growth.

A solution is neutral at pH 7, alkaline above pH 7, and acid below pH 7. The numbers are exponential: a solution of pH 6 has 10 times the hydrogen ion (H^+) content as a solution registering pH 7, a pH 5 solution has 10 times as many H^+ ions as a pH 6 solution, and so on. In the opposite direction, a solution at pH 8 has 10 times as many hydroxyl (OH^-) ions as a solution at pH 7, etc. The typical pH range for tissue culture media is between pH 4.5 and pH 5.7 (Figure 5-2).

In general, plant response is not too sensitive to changes in pH over a small range. Occasionally, however, a medium that has grown plant material for a period of time may develop a change in pH detrimental to the culture. In such cases, the culture should be transferred to a fresh medium, and the old medium should be checked to see if the pH might have been a cause of the culture's decline. If a significant change in pH has occurred it is a sign that the culture should be transferred more often.

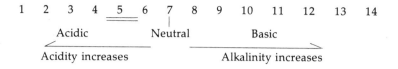

Figure 5-2. Range of pH.

When mixing formulas, a sodium hydroxide (NaOH) solution can be used to raise the pH of media, thus making the media more alkaline when they are too acidic. The NaOH solution contains sodium (Na^+) ions and hydroxyl (OH^-) ions. The OH^- ions of the NaOH solution combine with the excess H^+ ions in the medium to form water (H_2O), thus neutralizing the H^+ ions and causing the medium to be more basic (alkaline). Sodium hydroxide can usually be purchased in the form of pellets. Do *not* handle this chemical without using a forceps, spoon, or spatula—it is corrosive and will burn your skin. To make approximately a 1 *M* solution of NaOH, add 4 g of NaOH to 100 ml of water. Use great care when making this solution because when NaOH and water mix the reaction is violent and it may spatter. *Never* add water to NaOH; always add the NaOH to the water to minimize the caustic spatter that can injure eyes, skin, or anything it touches. You can also purchase a ready-made 1 *M* solution (sold as a 1 normal [1 *N*] solution) of NaOH. For NaOH and HCl, the normality and the molarity are the same.

Hydrochloric acid (HCl) solutions are used for lowering the pH of media, making them more acidic when they are too alkaline. The HCl solution contains hydrogen (H^+) ions and chloride (Cl^-) ions. The H^+ ions of the HCl solution combine with the excess OH^- ions in the medium to form water, thus neutralizing the OH^- ions and causing the medium to be more acidic. To lower the pH of media, use a 1 *M* solution of HCl. The molecular weight of HCl is 36.5, and it is often purchased as a 38% solution; therefore, 96 g of the 38% HCl solution is required for 1 liter of a 1 *M* HCl solution ($36.5 \div 0.38 = 96$ g HCl). To make 100 ml of 1 *M* HCl, tare (weigh) a beaker containing 75 ml of water. Then, using a pipet, slowly add the HCl until the balance indicates that 9.6 g of HCl have been added (weight of beaker and water plus 9.6 g HCl). *Never* add water to strong acid because it will spatter, harming anybody or anything it touches; instead, always add the acid to the water. After adding the HCl, remove the beaker from the balance and carefully add water to make 100 ml of solution. It may be easier to buy a 1 *N* solution of HCl already prepared.

The amount of sodium hydroxide (NaOH) or hydrochloric acid (HCl) required to change the pH one unit varies considerably, depending on the nature of the solution and at what point on the scale the change is made. The use of these NaOH and HCl solutions will be further considered in the chapter on making media (Chapter 6). More terms and concepts will be defined as we proceed, but the foregoing will be useful in the following discussion of various important chemicals. Not all the chemicals listed are used regularly in all media. Plants in culture manufacture many of the chemicals they need, and different plants have different requirements.

INORGANIC CHEMICALS

Inorganic chemicals and organic chemicals have different structural elements. Most inorganic chemicals have high melting points; few will burn; many are soluble in water; they conduct electricity in water; most are insoluble in organic solvents; and reactions involving inorganic chemicals tend to be very fast.

Growers will recognize immediately those essential elements that all plants require and so will not be surprised to find them in tissue culture media. Seven of these elements are major constituents in common fertilizers: calcium (Ca), iron (Fe), magnesium (Mg), nitrogen (N), phosphorus (P), potassium (K), and sulfur (S). The 3 remaining major elements—carbon (C), hydrogen (H), and oxygen (O)—are discussed under organic chemicals, in which they play a major role. It is also important now to recall the theory of limiting factors, discussed previously in Chapter 3. The German physical chemist Justus von Liebig theorized that if one essential element is missing, then all others present are of little or no value.

MAJOR ELEMENTS

Calcium (Ca), in the form of calcium pectate, is an integral component of plant cell walls, where it plays a role in the formation of pectin, a substance that bonds cell walls together. It helps control permeability and facilitates the movement of carbohydrates and amino acids throughout the plant; it also promotes root development. It assists in growth and development, as well as in nitrogen assimilation. As calcium oxalate, it ties up oxalic acid, which is a toxic by-product of protein metabolism. Dead shoot or root tips can be a sign of a lack of calcium. Calcium is usually included in tissue culture media as calcium chloride ($CaCl_2 \cdot 2H_2O$) or as calcium nitrate ($Ca[NO_3]_2 \cdot 4H_2O$); calcium phosphate (tribasic; $Ca_{10}[PO_4]_6[OH]_2$) is also sometimes included.

Iron (Fe) is involved in chlorophyll synthesis. It also participates in energy conversion in *photosynthesis* (the process by which carbon dioxide and water, with the help of chlorophyll and light, are converted into carbohydrates and oxygen is released) and respiration as it is reduced from the ferric (trivalent) to the ferrous (divalent) state. Plants deficient in iron have yellowed (*chlorotic*) young leaves, especially between veins or giving a striped appearance. In tissue culture media, ferrous sulfate ($FeSO_4 \cdot 7H_2O$) is often mixed with the sodium salt of ethylenediaminetetraacetic acid (Na_2EDTA; $C_{10}H_{16}N_2O_8Na_2 \cdot H_2O$) to sequester the iron, thereby making it more readily available to the plants. Iron is sometimes classified as a minor element.

Magnesium (Mg) is the central element in chlorophyll molecules. It is also important as an enzyme activator. Magnesium deficiency causes older leaves to become chlorotic. Most tissue culture formulas call for magnesium sulfate ($MgSO_4 \cdot 7H_2O$), commonly known as Epsom salts.

Nitrogen (N) influences the rate of plant growth. It is an essential element in the molecular make-up of nucleic acids, proteins, chlorophyll, amino acids, alkaloids, and some plant hormones. Lack of nitrogen is characterized by chlorotic leaves and stunted growth. Excess nitrogen promotes vigorous growth but suppresses fruit development. Common sources of nitrogen for tissue culture media are ammonium (NH_4^+) and nitrate (NO_3^-) compounds.

Phosphorus (P) is abundant in meristematic and other fast-growing tissues. As part of DNA and ATP (adenosine triphosphate) molecules, phosphorus is an essential element in photosynthesis and respiration, and it affects plant maturation and root growth. Stunted growth and reddish to purple coloring (*anthocyanin*) can be symptoms of phosphorus deficiency. Potassium phosphate (KH_2PO_4) and sodium phosphate ($NaH_2PO_4 \cdot H_2O$) are routinely included in tissue culture media.

Potassium (K) is necessary for normal cell division and promotes meristematic growth. It plays a role in many reactions within plants, although it is not an actual component of plant protoplasm, fats, or carbohydrates. It helps in synthesizing carbohydrates and proteins, manufacturing chlorophyll, and reducing nitrates. Insufficient potassium results in weak and abnormal plants, sometimes with mottled, curled, or dead leaf edges. Potassium nitrate (KNO_3) and potassium phosphate (KH_2PO_4) are common sources of potassium in culture media; potassium chloride (KCl) is used occasionally.

Sulfur (S) is present in some proteins. It promotes root development and deep green foliage. It is supplied in tissue culture media in sulfate (SO_4^{2-}) compounds.

MINOR ELEMENTS

In addition to the major elements that plants require, a number of other elements are essential to good growth but are needed only in extremely small quantities. They are called minor elements, trace elements, or micronutrients. As the purity of the water and the chemicals used in tissue culture media became more refined with the advances in technology, deficiency symptoms appeared in the media. This occurred because certain trace elements that had previously been supplied as undetected impurities in presumably pure water and chemicals were in fact necessary. Trace elements are present in soil, water, and even dust particles in adequate amounts to affect plant growth. Some "chemically pure" compounds that are used in media may contain traces of these elements. If so, they are usually listed as impurities on the label. Several trace elements are toxic to plants in excess amounts.

Boron (B) is an important trace element presumed to play a role in the movement of sugar, water, and hormones. It is also involved with nitrogen metabolism, fruiting, and cell division. Lack of boron produces interesting deficiency symptoms—often a deterioration of internal tissues, as in heart rot of sugar beets *(Beta vulgaris)*, cracked stem of celery *(Apium graveolens)*, or monkey face in olives *(Olea europaea)*. Other symptoms include tip die-back, or thick, curled, brittle, chlorotic leaves. Excessive amounts of boron can cause plant injury or death; thus some herbicides are borates. Boron is added to tissue culture media in small amounts as boric acid (H_3BO_3).

Chlorine (Cl) helps stimulate photosynthesis and seems to be necessary for growth. Deficiency symptoms are wilted leaves that become yellowed or bronze and die. Plants require chlorine in only minute quantities, but it is included in some tissue culture media in large amounts as calcium chloride $(CaCl_2 \cdot 2H_2O)$. These amounts appear to be tolerated by most plants, but caution is advised by some scientists who avoid including chlorine compounds in some formulas, choosing to get the calcium from sources other than calcium chloride.

Cobalt (Co) is an element in the complex vitamin B_{12} molecule and is essential to *nitrogen fixation* (the conversion of atmospheric nitrogen to nitrates by means of nitrogen-fixing bacteria). Cobalt chloride $(CoCl_2 \cdot 6H_2O)$ is added to most media in amounts of 0.025 mg per liter.

Copper (Cu) deficiency results in stunted growth, malformations, twisted and blotched leaves, or die-back of young twigs. Copper is believed to be necessary in energy conversion as it alternates between the cuprous (monovalent) and the cupric (divalent) state. It is involved with chlorophyll synthesis and is found in some enzymes. Only 0.025 mg of cupric sulfate $(CuSO_4 \cdot 5H_2O)$ per liter of medium is required to supply the necessary copper in tissue culture of most plants.

Iodine (I) is often added to media as potassium iodide (KI). It is not usually considered an essential element, even though it is a component of some amino acids. Iodine appears to have an adverse affect on *Rhododendron* cultures, so it should be omitted from that medium.

Manganese (Mn) deficiency is characterized by mottled yellowing of leaves. It is an essential element in chloroplast membrane. Manganese sulfate $(MnSO_4 \cdot H_2O)$ supplies the necessary manganese for tissue culture media.

Molybdenum (Mo) is believed to help convert nitrogen to ammonia and aids in nitrogen fixation. It is required for normal growth and protein synthesis. Its absence is suspect when leaves yellow between veins. Lack of molybdenum will cause whiptail (narrow leaf) condition in cauliflower *(Brassica oleracea)*, or it can stunt the growth of legumes (family Leguminosae). It is more available to plants in alkaline soils than in acid soils. Molybdenum is added to tissue culture media as sodium molybdate $(Na_2MoO_4 \cdot 2H_2O)$, also known as molybdic acid sodium salt. Quantities exceeding 10 parts per million can be injurious to plants.

Zinc (Zn) is an enzyme activator involved in chlorophyll formation, as well as in the production of the auxin indole-3-acetic acid (IAA). Without zinc, roots may be abnormal and leaves may turn mottled bronze or yellow and misshapen. A trace of zinc is included in tissue culture media as zinc sulfate ($ZnSO_4 \cdot 7H_2O$). Large quantities of zinc are toxic to plants, as is true of most trace elements.

ORGANIC CHEMICALS

In biology the most important molecules are organic molecules. Organic molecules are those molecules that contain at least one carbon atom, which is bonded (covalently, i.e., sharing electrons in pairs, with each atom providing half the electrons) to another carbon atom or to a hydrogen atom. Examples of organic compounds include carbohydrates, hormones, proteins, and enzymes. They can be gases, liquids, or solids with low melting points (less than 680°F [360°C]). They are mostly insoluble in water, and they do not conduct electricity in water. Most organic compounds can burn, and they usually react slowly.

Plants normally manufacture their own organic compounds, so they rarely need to be fed to outdoor or greenhouse plants. However, plants in culture cannot synthesize all the organic chemicals they need, so organic substances must be added to tissue culture media to augment the plantlets' self-generated (autotrophic) supply.

CARBOHYDRATES

Carbohydrates include such organic chemicals as sugars, starches, and cellulose. In varying amounts and configurations, carbon (C), hydrogen (H), and oxygen (O) are the primary elements that make up the molecules of carbohydrate compounds. These elements are generously supplied as carbon dioxide (CO_2) in air and as water (H_2O).

Sucrose ($C_{12}H_{22}O_{11}$), a disaccharide, is a common carbohydrate found in abundance in plant tissue. (A *polysaccharide* is a chain of one or more simple sugars linked together in its chemical structure.) Sucrose consists of 2 chemically bound monosaccharides, fructose ($C_6H_{12}O_6$) and glucose ($C_6H_{12}O_6$). Sucrose is an indirect product of photosynthesis. Plants growing in culture cannot manufacture all the sugar they require, so a high concentration of sucrose—generally 30 g per liter—is suggested for most media formulas. Sugar from sugar cane *(Saccharum officinarum)* or from sugar beet *(Beta vulgaris)*, both virtually 100% pure sucrose, as purchased from the grocery store are good sucrose for most tissue culture media. Occasionally fructose or glucose are substituted for sugar in media.

D-*Mannitol* is a sugar alcohol used as a nutrient and osmoticum (osmosis control), particularly when inducing plant protoplast formation and fusion.

D-*Sorbitol* is a sugar alcohol that is the primary translocatable (able to move through phloem and xylem tissue) carbohydrate in some plants. It is occasionally added to media.

VITAMINS

The vitamin B complex contains essential compounds for plant metabolism and growth. The growth substances found in yeast extract, which was commonly used in culture media in the past, are now identified as thiamine (vitamin B_1), nicotinic acid (niacin, or vitamin B_3), and pyridoxine (vitamin B_6), all of which are members of the vitamin B complex.

Adenine (vitamin B_4 or 6-aminopurine; $C_5H_5N_5$) is important to cells as part of the nuclear substances (DNA and RNA). It has a weak cytokinin effect. Adenine is used in culture media as adenine sulfate ($AdSO_4$; $[C_5H_5N_5]_2 \cdot H_2SO_4 \cdot 2H_2O$) to promote shoot formation.

d-Biotin (a B vitamin or vitamin H; $C_{10}H_{16}N_2O_3S$) is important in fat, protein, and carbohydrate metabolism. It is commonly used in media.

Choline ($C_5H_{15}NO_2$) is an *alkaloid* (a compound with alkaline properties) within the vitamin B complex. It occurs naturally in lecithin, which is chemically related to fats but also contains phosphorus and nitrogen. Choline chloride ($C_5H_{14}NOCl$) is occasionally specified in media formulas.

Cyanocobalamin (vitamin B_{12}; $C_{63}H_{88}CoN_{14}O_{14}P$) is sometimes added to culture media as a possible growth promoter.

Folic acid (vitamin Bc or vitamin M; $C_{19}H_{19}N_7O_6$) is found in leaves and other plant tissues. It functions as a B vitamin and demonstrates *coenzyme* (an organic molecule associated with enzymes) activity.

Inositol (myo-inositol; $C_6H_{12}O_6$), a sugar alcohol in the B complex, is required in many media. In its phosphate form it is part of various membranes, particularly those of organelles such as chloroplasts. Inositol, though not essential to growth, is often beneficial when added to tissue culture media at 100 mg per liter.

Nicotinic acid (niacin or vitamin B_3; $C_6H_5NO_2$) is a component of coenzymes active in light-energy reactions. Various media require nicotinic acid, usually in amounts ranging from 0.1 to 10 mg per liter.

PABA (*para*-aminobenzoic acid, vitamin B_x; $C_7H_7NO_2$), occasionally used in culture media, serves as an antiseptic and a preservative. It also plays a role in folic acid metabolism.

D-Pantothenic acid (vitamin B_5; $C_9H_{16}NO_5$) is active as a coenzyme in fat metabolism. It is added as the calcium salt ($[C_9H_{16}NO_5]_2Ca$) to some plant tissue culture media. It should be cold sterilized.

Pyridoxine (vitamin B_6; $C_8H_{11}NO_3$) also serves as a coenzyme in some metabolic pathways (chemical reactions of metabolism). It is usually included in culture media as the hydrochloride form ($C_8H_{11}NO_3 \cdot HCl$).

Riboflavin (vitamin B_2 or vitamin G; $C_{17}H_{20}N_4O_6$) is active in carbohydrate metabolism and is essential to cellular respiration.

Thiamine (vitamin B_1; $C_{12}H_{17}ClN_4OS$) is essential for most culture media because it functions as a coenzyme to assist the organic acid cycle of respiration (known as citric acid cycle or Krebs cycle). Only 0.4 mg of thiamine hydrochloride per liter is specified for many tissue culture media.

L-Ascorbic acid (vitamin C; $C_6H_8O_6$) has some disinfectant qualities, but its chief use in tissue culture is as an antioxidant to prevent *phenolic oxidation* (browning of plants that contain phenolics). It should not be used for extended periods because it can become an oxidant itself.

(+)-α-Tocopherol (vitamin E; $C_{29}H_{50}O_2$) is occasionally used in culture media. It is known to promote dispersion in suspension cultures of ladino clover (*Trifolium repens* 'Royal Ladino') and soybean *(Glycine max)* (Oswald et al. 1977). Later studies showed that cell aggregation was increased when vitamin E was removed from the medium.

GROWTH REGULATORS

Growth regulators, or hormones, are not nutrients, but they influence growth and development. They are generally produced naturally in plants. Cultures, however, usually do not manufacture sufficient quantities of growth regulators, so they must be added selectively to culture media. Auxins, cytokinins, gibberellins, and ethylene are among the substances used as hormones or growth regulators. Generally speaking, auxins promote cell enlargement and root initiation, whereas cytokinins promote cell division and shoot initiation. This is a simplistic summary in view of the diversity of growth regulators and their effects. There is a wide range of interactions between auxins and cytokinins. They also interact with other chemicals and are affected by environmental factors, such as light and temperature. Under some conditions an auxin might even react as a cytokinin and a cytokinin as an auxin. It is important that tissue culture media contain the right kinds of hormones and in the correct ratio for each variety of plant being cultured. Occasionally cultures become *habituated*, a state in which they lose the need for the hormones in the medium or simply do not respond to them.

Auxins

Auxins are phytohormones that influence cell enlargement, root initiation, and adventitious bud formation. They suppress the initiation of lateral buds (which is the bud of choice for ensuring genetic stability). Auxins are commonly used in tissue culture media, either combined with cytokinins during the multiplication stage (Stage II) or without cytokinins for the rooting stage (Stage III). Dissolve auxins in a few drops of 1 M NaOH or 1 M potassium hydroxide (KOH) before adding it to media.

Dicamba (3,6-dichloro-o-anisic acid or 3,6-dichloro-2-methoxybenzoic acid; $C_8H_6Cl_2O_3$) is a synthetic, auxin-type growth regulator that is used as an herbicide. It is used in some media to promote callus. Dissolve dicamba in 1 M NaOH.

2,4-D (2,4-dichlorophenoxyacetic acid; $C_8H_6Cl_2O_3$) is a synthetic auxin known primarily as a weed killer. It has been widely used in plant tissue culture media to induce callus growth.

IAA (indole-3-acetic acid; $C_{10}H_9NO_2$) is best known to plant propagators as a dip to promote rooting of cuttings, and it is often used for rooting in tissue culture media. It is unstable in light, so it should be stored in darkness.

IBA (indole-3-butyric acid; $C_{12}H_{13}NO_2$) is another rooting hormone commonly used by propagators. It is produced naturally in plants and also made synthetically. More stable than IAA, IBA is the preferred hormone for root induction in tissue culture.

NAA (1-naphthaleneacetic acid; $C_{12}H_{10}O_2$) is another root-inducing, auxin-type compound that is sometimes used in media, especially to promote callus growth.

Cytokinins

Cytokinins, formerly called kinins, are growth regulators that are required in tissue culture media for cell division, shoot multiplication, and axillary bud proliferation. They help delay *senescence* (aging), and they influence auxin transport. If cultures are too spindly, increased cytokinin will help foster shorter, stouter stems. Cytokinins dissolve in about 1 M HCl. They are usually omitted from media for the rooting stage.

BA or *BAP* (6-benzylaminopurine; $C_{12}H_{11}N_5$) is probably used more often than 2iP in culture media to promote axillary bud growth. It is usually made synthetically; although this cytokinin is presumed to be common in plants, it is difficult to find. 6-Benzylaminopurine riboside ($C_{17}H_{19}N_5O_4$) is often used as an alternative to BA.

BPA or *PBA* (N-benzyl-9-[2-tetrahydropyranyl]adenine or 6-benzylamino-9-[2-tetrahydropyranyl]-9H-purine; $C_{17}H_{19}N_5O$) is a synthetic cytokinin occasionally used in cultures for axillary bud and callus proliferation.

2iP (IPA; 6-[γ,γ-dimethylallylamino]purine or N^6-[2-isopentenyl]adenine; $C_{10}H_{13}N_5$) is commonly used in plant tissue culture media. It is made synthetically, but it has been found in RNA and in a pathogenic bacterium *(Corynebacterium fasscians)*. It causes rapid cell division and consequent irregular growth in some higher plants.

Kinetin (6-furfurylaminopurine; $C_{10}H_9N_5O$) is a growth regulator that is isolated from DNA. It is often used in plant tissue culture media to promote cell division.

Zeatin (6-[4-hydroxy-3-methylbut-2-enlyamino]purine; $C_{10}H_{13}N_5O$) was first discovered in corn seed endosperm. It is a common alternative cytokinin to 2iP or BA for use in media.

Gibberellins

Gibberellins are a group of naturally occurring substances that influence cell enlargement and stem elongation. Kurasawa noted in 1926 that secretions from a fungus *(Gibberella fujikuroi)* resulted in abnormally rapid growth in rice *(Oryza)* seedlings. The substance was gibberellic acid (GA$_3$; C$_{19}$H$_{22}$O$_6$), which was later isolated in crystalline state from both fungi and higher plants. Thirty-four gibberellins have been chemically identified. Some of them appear in embryos, where they initiate production of the enzyme alpha amylase, which converts starches to sugars and stimulates other enzymes. GA$_3$ is sometimes used as a growth regulator to supplement auxins and cytokinins.

Other growth-regulating substances

Ancymidol (C$_{15}$H$_{16}$N$_2$O$_2$), available commercially as A-Rest, has a growth-retardant, cytokinin effect in tissue culture. It has been found to shorten shoots and induce buds in some cultures.

CPA (para-chlorophenoxyacetic acid; C$_8$H$_7$ClO$_3$) is an auxin-type growth regulator.

Ethylene (C$_2$H$_4$) is a gas produced by plants. It has growth regulatory properties and is involved with fruit ripening, flowering, and leaf abscission. Its build-up in culture vessels can be detrimental, however. Ethylene production is greater than normal in vitrified cultures (Kevers and Gaspar 1985). (*Vitrification* is a phenomenon in which tissues develop a glassy, swollen appearance.)

Paclobutrazol (DL-1,2,4-triazole-3-alanine) is useful as an addition to Stage III media to help minimize the need for acclimatization (Oliphant 1990; Smith et al. 1992). It is a growth-retardant substance that has a cytokinin effect in many tissue cultures.

Phloroglucinol (1,3,5-trihydroxybenzene) is sometimes used as an antioxidant (Smith 1992) and in the treatment of vitrification (Phan and Hagadus 1986). It is also somewhat bactericidal and sometimes promotes culture growth (Donnelly and Vidaver 1988).

TIBA (2,3,5-triiodobenzoic acid), an antiauxin, inhibits auxin movement and may be growth promoting in culture. It also may inhibit unwanted callus.

AMINO ACIDS

Amino acids are building blocks of proteins, some of which combine with nucleic acids to form nucleoproteins. Although many more can be synthesized in laboratories, only 20 essential amino acids are found in nature. These are all in the L chemical form. Several are particularly pertinent to cell culture. Others are only occasionally called for in certain media.

L-Alanine (C$_3$H$_7$NO$_2$) has helped to increase the number of embryos in some cell cultures.

L-Arginine (C$_6$H$_{14}$N$_4$O$_2$) reportedly has assisted root initiation.

L-Asparagine (C$_4$H$_8$N$_2$O$_3$) is sometimes used in cell culture of soybean *(Glycine max)*. When combined with proline, L-asparagine and certain other amino acids assist in embryogenesis in alfalfa *(Medicago sativa)*.

L-Cysteine (C$_3$H$_7$NO$_2$S), a sulfur-containing amino acid, is occasionally called for in media formulas. It is added to media as the hydrochloride (C$_3$H$_7$NO$_2$S · HCl).

L-Glutamine (L-2-aminoglutaramic acid; C$_5$H$_{10}$N$_2$O$_3$) can contribute to somatic embryogenesis.

Glycine (aminoacetic acid; C$_2$H$_5$NO$_2$) is another amino acid specified in some media. It is sometimes used in *Begonia* media and has been used in a medium for *Poinsettia* cell culture.

L-Lysine (L-2,6-diaminohexanoic acid; C$_6$H$_{14}$N$_2$O$_2$) combined with proline has enhanced the quality and number of embryos produced.

L-Proline (C$_5$H$_9$NO$_2$) is often used in combination with other amino acids for embryo culture.

L-Serine ($C_3H_7NO_3$) has been used in microspore (pollen) culture to grow haploid embryos.

L-Tyrosine (L-3-[4-hydroxyphenyl]alanine; $C_9H_{11}NO_3$) is effective in shoot initiation and is useful as a nitrogen source.

ANTIBIOTICS

An *antibiotic* is a substance produced by plants or microorganisms (or made synthetically) that has a toxic effect on other microorganisms; antibiotics can retard or prevent the growth of such microorganisms, or kill them. Antibiotics are not used routinely against contaminants in micropropagation because often they are ineffective, kill the culture, or induce chromosomal instability. However, some have been useful on occasion. More commonly, antibiotics are used in gene-transfer experiments to control or eliminate *Agrobacterium* (Smith 1992).

Ampicillin has been used in amounts of 100 to 400 mg per liter in culture of *Petunia* protoplasts (Dixon 1985).

Carbenicillin (at 500 mg per liter) and *augmentin* (at 250 mg per liter) were used advantageously in some genetic-engineering experiments for controlling *Agrobacterium* (Smith 1992).

Cefotaxime has been used to eliminate certain bacteria in woody plants. It has been used at 25 mg per liter in combination with *tetracycline* (at 6 mg per liter), *rifampicin* (at 6 mg per liter) dissolved in dimethyl sulfoxide (DMSO; C_2H_6OS), and *polymyxin B* (at 25 mg per liter). These should be cold sterilized because they degrade when heat treated.

Gentamicin sulfate is described as an autoclavable antibacterial. The Sigma Chemical Company suggests its use at 50 mg per liter. It has been used in enzyme solutions for *Petunia* protoplasts at 10 mg per liter, combined with ampicillin (400 mg per liter) and tetracycline (10 mg per liter) (Dixon 1985).

Methylolurea is a useful antibiotic against certain yeasts. It is available commercially as methylolurea solution or UF-85. It is a fertilizer with nitrogen derived from products of urea-formaldehyde reactions. Most antibiotics do not withstand heat sterilizing, but this one is autoclavable.

Polymyxin B is an antibiotic occasionally used in cell culture.

Ribavirin (virazole; 1-β-D-ribofuranosyl-1,2,4-triazole-3-carboxamide) is a broad-spectrum antiviral agent that is particularly effective against some potato viruses.

Streptomycin, used at 20 to 100 mg per liter, may be useful even if autoclaved.

Other antibiotics, such as *bacitracin, mycostatin, nystatin, penicillin, phosphomycin,* and *terramycin,* have been tried in media, but with little or no success.

CHEMICALLY UNDEFINED CONSTITUENTS OF TISSUE CULTURE MEDIA

Agar is a mixture of polysaccharides derived from extracts of several species of red algae. As a gelling agent, the agar in tissue culture media is strong enough to support the culture, yet liquid enough to allow the nutrients to diffuse through the medium to the plantlets. Although agar is supposedly an inert material, in tissue culture media it frequently contains traces of other elements. Because it is not completely defined chemically, any medium containing true agar (as opposed to synthetic agar) must be considered only a partially defined medium. Agar may be found on the market as Agar-Agar or Gum Agar.

The concentration of impurities in agar varies with the source of the algae and the method of manufacture. Greater trace amounts of chloride were found in agar produced from Spanish sources than in those from Japanese and Portuguese sources, for example. Other impurities discovered in some agars include sulfate, calcium, magnesium, and iron ions (Rechcigl 1978). Purified agars sold by tissue culture suppliers are usually sufficiently pure for plant tissue culture.

Six to eight grams of agar per liter of medium is usually satisfactory, but gel strength will vary both with the medium formula being used and the source and grade of agar. Gel strength was found to vary by 35% between the lowest and the highest gels in a group of agar samples from different sources. The gel should be firm enough that test tubes will hold a good slant without being sloppy, yet soft enough that plant material can be pressed gently and easily for good contact with the agar. Agar media with low salts or hormones will tend to be firmer than those with high salt and hormone contents. Media with a low pH (around pH 4.5) will tend to be softer than media with a higher pH (pH 5.7). To some extent, depending on the plant, gel strength can affect plant performance. Hard agar may slow nutrient movement through the medium. Soft agar may cause vitrification.

Several agar substitutes, which are highly refined polysaccharides, are commercially available. Trade names for such agar substitutes include Gelrite, Gellan, and Phytagel. Media made with these substitutes are exceedingly clear and usually have no ill effects, but they may contribute to vitrification. Agargel (a blend of agar and Phytagel) is one of several blends on the market; it should be used at 3.5 to 5 g per liter. We prefer to mix Gelrite (or equivalent) with agar in ratios of 3:1 to 5:1 of Gelrite to agar, depending on the desired strength. The cost is less than it is for general-purpose agar alone because less Gelrite is required to provide the same gel strength.

Activated charcoal is frequently added to rooting media to adsorb root-inhibiting agents. Sometimes it is added to Stage I media to adsorb toxic phenolics, and charcoal has also been included in Stage II media intermittently for curative purposes. It is usually added at 0.6 g per liter.

Casein hydrolysate (edamin) is an undefined protein mixture that is occasionally used in media as a non-specific source of organic nitrogen. Casein, a phosphoprotein in milk, is hydrolyzed (treated with water) to form this weak acid.

Coconut milk is the liquid endosperm of coconut *(Cocos nucifera)*. This undefined medium was used with success in early plant tissue culture and is still used today. It was first reported in 1941 by J. van Overbeek in tissue culture of jimson weed *(Datura)* embryos. In 1948, Caplin and Steward grew carrot phloem explants using coconut milk and casein hydrolysate as basal salt supplements. Steward achieved 80-fold multiplication of carrot roots in 3 weeks using coconut milk. Coconut milk has also been used widely in orchid culture. In the early 1960s Carl Withner first cultured *Phalaenopsis* orchids on Vacin and Went's medium, which contains 15% coconut milk.

Coconut milk is available as coconut water from many commercial suppliers, or you can extract it yourself from a coconut. To do this, obtain a green coconut that has reached full size (as young or fresh as you can find in the store), using the heaviest one you can find. Shake the coconut before you buy it to make sure there is liquid in it—you will hear it sloshing. Pound 2 holes in the eyes with a clean nail. Pour out the liquid. Filter it through filter paper and store in the freezer.

Corn milk (corn endosperm) is a rare additive to tissue culture media. It is suggested that the presence of the cytokinin zeatin in corn milk may be the source of benefit.

Cornstarch has only recently been reported as an agar substitute (Stanley 1995). Ingrid Fordham of the U.S. Department of Agriculture's Agricultural Research Service Fruit Laboratory has found that it is as effective or better than agar as a substrate for certain apples *(Malus)*, berries, and pears *(Pyrus)*. She recommends 50 g of cornstarch, plus 0.5 g of Gelrite, for 1 liter of medium. At a 99% savings compared to agar, cornstarch is sure to find wider usage.

Potato extract is commonly added to media for monocot and anther culture.

Yeast extract is a natural source of vitamins. It is purchased as a powder and is occasionally used in media.

CHEMICAL LIST

Because the total number of chemicals required for tissue culture media can be somewhat overwhelming, the list of chemicals is divided into convenient reference sections. Some of the chemicals in the list are seldom required; they are included for those who wish to experiment or for the few occasions in which they are prescribed. Only those chemicals commonly used in culture media should be stocked initially; for example, those used in MS medium, McCown and Lloyd's woody plant medium (WPM; see Section II, *Kalmia*), or Anderson's medium (see Section II, *Rhododendron*). Other important compounds that you will encounter in the various media formulas in Section II are also listed, and some others that come up in Section II are not listed here. At first, do not buy more chemicals than you know you will need for particular media.

Compound	Amount to buy	Cost (US$)
INORGANIC CHEMICALS FREQUENTLY USED		
Ammonium nitrate (NH_4NO_3)	500 g	18.95
Ammonium sulfate ($[NH_4]_2SO_4$)	500 g	28.85
Boric acid (H_3BO_3)	100 g	8.40
Calcium chloride ($CaCl_2 \cdot 2H_2O$)	500 g	26.90
Calcium nitrate ($Ca[NO_3]_2 \cdot 4H_2O$)	500 g	26.10
Cobalt chloride ($CoCl_2 \cdot 6H_2O$)	25 g	9.65
Cupric sulfate ($CuSO_4 \cdot 5H_2O$)	250 g	14.00
Ethylenediaminetetraacetic acid, disodium salt (Na_2EDTA)	100 g	15.65
Ferrous sulfate ($FeSO_4 \cdot 7H_2O$)	50 g	4.15
Magnesium sulfate ($MgSO_4 \cdot 7H_2O$)	500 g	18.25
Manganese sulfate ($MnSO_4 \cdot H_2O$)	100 g	11.80
Potassium iodide (KI)	100 g	21.70
Potassium nitrate (KNO_3)	500 g	23.80
Potassium phosphate (monobasic, anhydrous; KH_2PO_4)	100 g	11.10
Sodium molybdate (dihydrous; $Na_2MoO_4 \cdot 2H_2O$)	100 g	22.60
Sodium phosphate (NaH_2PO_4)	100 g	11.80
Zinc sulfate ($ZnSO_4 \cdot 7H_2O$)	100 g	13.35
ORGANIC CHEMICALS FREQUENTLY USED		
Adenine sulfate ($C_5H_5O_5 \cdot H_2SO_4 \cdot 2H_2O$)	5.0 g	10.55
Agar (Agar-Agar or Gum Agar)	1.0 kg	96.35
Agar blend (Agargel or equivalent)	500 g	57.25
Agar substitute (Phytagel or equivalent)	1.0 kg	128.50
6-Benzylaminopurine (BA, BAP)	500 mg	8.10
Charcoal, activated	500 g	14.70
6-(γ,γ-Dimethylallyamino)purine (2iP)	1.0 g	37.00
Indole-3-acetic acid (IAA)	5.0 g	9.15
Indole-3-butyric acid (IBA)	1.0 g	6.30
myo-Inositol	50 g	12.00
Kinetin (6-furfurylaminopurine)	100 mg	6.80
Nicotinic acid (niacin, vitamin B_3)	25 g	4.85
Pyridoxine HCl (vitamin B_6)	25 g	17.85
Sucrose (sugar)	4.5 kg (10 lb)	3.00
Thiamine HCl (vitamin B_1)	5.0 g	5.55

Compound	Amount to buy	Cost (US$)
COMPOUNDS USED INFREQUENTLY		
L-Alanine	100 mg	6.00
para-Aminobenzoic acid (PABA)	1.0 g	6.90
Ancymidol (A-Rest)	25 mg	14.95
L-Arginine HCl	25 g	8.10
L-Ascorbic acid (vitamin C)	25 g	9.15
6-Benzylaminopurine riboside (alternative to BA)	100 mg	10.60
N-Benzyl-9-(2-tetrahydropyranyl)adenine (BPA, PBA)	10 mg	8.30
d-Biotin (vitamin H)	100 mg	10.10
Calcium phosphate (tribasic; $Ca_{10}[PO_4]_6[OH]_2$)	25 g	19.75
Carbenicillin	250 mg	24.50
Casein hydrolysate	250 g	16.60
Cefotaxime	100 mg	17.00
para-Chlorophenoxyacetic acid (CPA)	25 g	5.70
Choline chloride ($C_5H_{14}NOCl$)	5.0 g	18.50
Citric acid	100 g	10.10
Coconut milk	1 fruit	1.00
or commercial coconut water	100 ml	12.45
Cyanocobalamin (vitamin B_{12})	100 g	9.85
L-Cysteine HCl	5.0 g	7.30
2,4-Dichlorophenoxyacetic acid (2,4-D)	100 g	15.65
Ferric tartrate ($Fe_2[C_4H_4O_6]_3$)	100 g	14.35
Folic acid (vitamin Bc)	1.0 g	7.30
Gentamicin sulfate	50 mg	8.05
Gibberellic acid (GA_3)	500 mg	13.90
L-Glutamine	25 g	12.55
Glycine (aminoacetic acid)	100 g	11.75
L-Lysine	5.0 g	14.90
1-Naphthaleneacetic acid (NAA)	25 g	9.35
D-Pantothenic acid (vitamin B_5)	5.0 g	7.20
Phloroglucinol	25 g	7.70
Potassium chloride (KCl)	250 g	9.20
L-Proline	25 g	14.25
Riboflavin (vitamin B_2)	5.0 g	6.65
L-Serine	25 g	18.50
Streptomycin	5.0 g	8.20
Thidiazuron (1-phenyl-3-[1,2,3-thiadiazol-5-yl]urea)	25 mg	40.80
2-Thiouracil (4-hydroxy-2-mercaptopyrimidine;		
an antiviral agent)	25 g	8.65
(+)-α-Tocopherol (vitamin E)	10 g	6.20
L-Tyrosine	50 g	14.70
Yeast extract	100 g	19.15
Zeatin	5.0 mg	18.30
NON-MEDIA CHEMICALS		
Alcohol, ethyl	1 pt	30.00
Alcohol, isopropyl (70%)	1 pt	0.75
Hydrochloric acid (HCl; 1.0 M for pH adjustment)	1 gal	14.00
Hydrogen peroxide (3%; H_2O_2)	1 pt	0.75
Sodium hydroxide (NaOH; 1.0 M for pH adjustment)	1 gal	14.00
Sodium hypochlorite (bleach)	1 gal	1.00
Tween 20 (polyoxyethylenesorbitan, monolaurate;		
wetting agent)	100 ml	9.50

QUICK AND EASY

For the person who thinks tissue culture is all too complicated and "scientific," M. Bridgen (1986) proposed the ultimate do-it-yourself strategy, in which the grocery store, pharmacy, and health food store provide the sources for all the ingredients. They suggest the following off-the-shelf recipe:

Table sugar	⅛ cup
Tap water	1 cup
Nutrient solution: ¼ tsp all-purpose 10-10-10 fertilizer in 1 gal. water	½ cup
Inositol tablet (250 mg)	½ tablet
Vitamin tablet with thiamine	¼ tablet
Agar flakes	2 tablespoons

Combine the ingredients in a flask or beaker. Boil, while stirring, until the agar has melted. Dispense medium to about 0.5 in (or 1 cm) deep into pint canning jars or baby-food jars. Cover and process in a pressure cooker, according to the directions of the cooker manufacturer, for 15 minutes at 15 pounds pressure. Tweezers and razor blades can be sterilized in the cooker at the same time; wrap them in aluminum foil before placing in the cooker. At the same time, sterilize pint jars of water to use for explant cleaning.

Bridgen successfully propagated Boston fern (*Nephrolepis exaltata*) rhizome tips, African violet (*Saintpaulia*) leaf and petiole sections, and wandering Jew (*Tradescantia fluminensis*) shoot tips with this medium. Before proceeding, see Chapters 7 and 8 for information on explant preparation and sterile technique, and Section II for guidelines on particular plants. For the fun of it, experiment with other explants or ingredients. What happens if you substitute coconut milk for some of the tap water?

Chapter 6: Media Preparation

This chapter presents, in detail, the procedures you will want to follow for mixing media. They are written primarily for the novice, but those familiar with general laboratory practices may also find these descriptions useful.

It is important to consistently use caution, care, and common sense. Work slowly, especially at first, in order to avoid costly mistakes, such as spilling or wasting chemicals, using the wrong ingredients, or burning yourself. With practice you will become familiar with laboratory language and concepts, you will learn what to expect of your equipment and how to use it, and you will come to develop a special respect for and knowledge of the various chemicals. The concentration of salts, for example, particularly the level of ammonium ions (NH_4^+), can have a profound affect on culture performance. Reducing ammonium nitrate (NH_4NO_3) and potassium nitrate (KNO_3) levels in Murashige and Skoog's (MS) formula to one-quarter strength has been found to be beneficial in some cultures of Douglas firs *(Pseudotsuga menziesii)*, *Rhododendron*, tomatoes *(Lycopersicon)*, *Arabidopsis*, *Torenia*, and orchids, to mention a few.

For about US $25 you can purchase enough premixed powdered medium (without agar) to make 10 liters. Follow the directions accompanying the package. Except for one-time projects, most serious beginners will prefer to purchase the individual chemicals required.

The media formulas for *Begonia rex* (see Table 6-1) is an example of the formulas you will find in the literature, or later in Section II of this book. As you work through the procedures, the concepts will become more meaningful. *Begonia rex*, a common houseplant, is the example used for the micropropagation procedures described in this and subsequent chapters because it is easy to tissue culture and is commonly available where houseplants are sold. Although begonias might seldom be tissue cultured because of the ease of propagating them by cuttings, for efficient mass production or for reproduction of an exceptional individual plant, tissue culture of *Begonia* is an effective procedure. In general, if a plant is easy to propagate by conventional cuttings, it usually will respond well and easily to tissue culture.

STOCK SOLUTIONS

Stock solutions are concentrated solutions of groups of media chemicals that are prepared ahead of time and used to make several batches of media. They may be made in liter quantities of 10 or 100 times the concentration required in the final formula (below we are making concentrations of 100 times). Having stock solutions eliminates the need to weigh so many different chemicals every time you want to make a batch of medium. Also, the quantities will be more accurate because they are on a larger scale than would be required for a single batch of medium, and thus minor inaccuracies have less impact.

Some of the ingredients will *precipitate* (form insoluble compounds) if mixed together in concentrated form, so each group is made up of chemicals that usually will not precipitate at the concentration of the stocks. Before adding any chemicals in making the stock solutions, there must be water in the flask, otherwise precipitates are likely to form.

There is not total agreement on the combinations of stock ingredients, nor on how or for how long they can be stored. Most stocks can be stored for a limited time without adverse reactions. If they have a short shelf life, as do organics, then the chemicals will be stable for a longer time if they are stored in a refrigerator; hormones tend to have a particularly short shelf life—

which is why they are made here in small amounts of 25 mg per 250 ml of water. If the stock ingredients have a longer shelf life, as do the inorganic salts, then they can be stored in a cupboard, but they run a greater risk of growing contaminants because of the warmer temperature there. On the other hand, if the salt solutions are bordering on forming precipitates, they will do so in the cold of the refrigerator because the solubility decreases with decreases in temperature. If solutions form precipitates, they can be brought up to room temperature, or heated further, and then used, providing the precipitates dissolve. If the stock does not go into solution easily or precipitates, heating it on a hot plate/stirrer will often solve the problem.

If you spill some chemicals or remove more than you need, it is bad practice to return the excess to the bottle from which it came because it may have mixed with other substances. It is also important that you be sure to clean the spoon or spatula between weighing different chemicals. A fine paint brush is handy for brushing the last bits of chemical from the weighing paper into the flask.

MS STOCK SOLUTIONS FOR BEGONIA REX MEDIA

Equipment and supplies
    1 1-liter Erlenmeyer flask
    2 400-ml beakers
    Distilled water
    Hot plate/stirrer and stir bar
    2 balances (one for small measurements and one for measurements of more than
       1 g)
    Weighing papers
    1 spatula
    Forceps
    Wash bottle containing distilled water
    Medicine dropper
    1 M NaOH (to dissolve auxins; see Chapter 5)
    1 M HCl (to dissolve cytokinins; see Chapter 5)
    Containers, with lids (for storing stock solutions)

MS salts stocks
    Nitrate stock (1 liter)
       1. Pour 700 ml of distilled water into a 1-liter flask.
       2. Slide the magnetic stir bar into the flask (see Figure 4-6).
       3. Place the flask on the magnetic stirrer and turn on the stirrer.
       4. Place a weighing paper (or piece of aluminum foil) on the balance and tare (weigh) it, then set the balance for 190 g more than the weighing paper.
       5. Using a spatula (or spoon or knife) weigh:
           Potassium nitrate (KNO_3)   190 g
       6. Pour the KNO_3 into the flask on the stir plate.
       7. Place a weighing paper on the balance and tare it, then set the balance for 165 g more than the weighing paper.
       8. Weigh and add:
           Ammonium nitrate (NH_4NO_3)   165 g
           Total   355 g
       9. Stir until dissolved.

10. Turn off the stirrer and remove the stir bar with long forceps, or pour the stock solution into a clean flask, saving out the stir bar.

11. Add distilled water to bring the level of solution up to 1000 ml (1 liter).

In a similar manner make the remaining stocks:

Sulfate stock (1 liter)

Magnesium sulfate ($MgSO_4 \cdot 7H_2O$)	37.0 g
Manganese sulfate ($MnSO_4 \cdot H_2O$)	1.69 g (1690 mg)
Zinc sulfate ($ZnSO_4 \cdot 7H_2O$)	0.86 g (860 mg)
Cupric sulfate ($CuSO_4 \cdot 5H_2O$)*	0.0025 g (2.5 mg)
Total	39.6 g

Halide stock (1 liter)

Calcium chloride ($CaCl_2 \cdot 2H_2O$)	44.0 g
Potassium iodide (KI) (often omitted)	0.083 g (83 mg)
Cobalt chloride ($CoCl_2 \cdot 6H_2O$)*	0.0025 g (2.5 mg)
Total	44.1 g

Phosphate stock (1 liter)

Potassium phosphate (KH_2PO_4)	17.0 g
Boric acid (H_3BO_3)	0.62 g (620 mg)
Sodium molybdate ($Na_2MoO_4 \cdot 2H_2O$)	0.025 g (25 mg)
Total	17.6 g

Iron stock (1 liter)

Iron stock should be heated until dissolved and stored in the dark.

Ferrous sulfate ($FeSO_4 \cdot 7H_2O$)	2.78 g
EDTA disodium salt (Na_2EDTA)	3.73 g
Total	6.5 g

Total MS salts in these 5 stock solutions: 462.8 g.

Vitamin stocks (store in refrigerator)

Inositol/Thiamine stock (1 liter)

Inositol	10.0 g
Thiamine HCl	0.04 g (40 mg)
Total	10.04 g

Nicotinic acid/Pyridoxine stock (1 liter)

Nicotinic acid	0.1 g (100 mg)
Pyridoxine HCl	0.1 g (100 mg)
Total	0.2 g (200 mg)

Auxin stock (250 ml)

Auxin stock should be stored in a refrigerator. This stock contains 0.1 mg of 1-naphthaleneacetic acid (NAA) per milliliter of solution. Stock solutions of other auxins, such as IAA and IBA, can be made in the same way as described here for NAA.

*Cobalt chloride and cupric sulfate are required in amounts (2.5 mg) that are too small to weigh on many balances. To obtain this amount, weigh 25 mg of cupric sulfate and add it to 100 ml of water. Using a pipet, take 10 ml of this solution to provide 2.5 mg of cupric sulfate. Do the same for cobalt chloride.

1. Weigh 25 mg of NAA and place in a 400-ml beaker.
2. Using a dropper, slowly add, while stirring with a spatula, several drops of 1 *M* NaOH or KOH until the NAA crystals are dissolved.
3. Quickly add 250 ml of distilled water.
4. Transfer solution to a container and close tightly.

Cytokinin stock (250 ml)

Cytokinin stock should be stored in a refrigerator. This stock contains 0.1 mg of 6-benzyl-aminopurine (BA) per milliliter of solution. Stock solutions of other cytokinins, such as 2iP and kinetin, can be made in the same way as described here for BA.

1. Weigh 25 mg of BA and place in a 400-ml beaker.
2. While stirring, add 3 or 4 drops of distilled water.
3. Add 1 *M* HCl solution one drop at a time until the BA is dissolved. (Apply a little heat to help dissolve the crystals.)
4. Quickly add 250 ml of distilled water.
5. Transfer solution to a container and close tightly.

MEDIA MIXING

Having made the stock solutions, you are at last ready to mix the medium, a task much easier than mixing stocks.

Table 6-1. *Begonia rex* **Media**

Compound	Stages I & II	Stage III
	mg/liter	
MS salts	4,628	4,628
Inositol	100	100
Thiamine HCl	1.5	1.5
Nicotinic acid	0.5	0.5
Pyridoxine HCl	0.5	0.5
BA	0.4	–
NAA	0.1	0.1
Glycine (optional)	2.0	2.0
Sucrose	30,000	20,000
Agar	8,000	8,000

Calculating amount of stock solution per liter of medium

Although the calculations for the amount of stock solutions needed have already been done for the *Begonia* medium in progress (see the checklist in Table 6-2), it would be good practice for you to figure for yourself how the amounts are determined. To determine the amount of stock solutions required for a medium, the easiest arithmetic to use is simple proportion.

You already have made a cytokinin stock solution with 25 mg of BA in 250 ml of water. Say for a particular medium you require only 0.4 mg of BA, so you need to know how many milliliters of stock solution to use in order to obtain that amount of BA. Write the original milligram amount of BA (25 mg) over the original milliliter total of stock solution (250 ml). Note: it does not matter if you have already used some from the stock solution, the *proportion* is still the same. Next, write the milligram amount you need (0.4 mg) over the unknown milliliter amount (? ml):

$$\frac{25 \text{ mg}}{250 \text{ ml}} = \frac{0.4 \text{ mg}}{? \text{ ml}}$$

Cross multiply:

$$25 \times ? = 250 \times 0.4$$
$$25 \times ? = 100$$

Divide both sides by 25 so the unknown will be by itself.

$$\frac{25 \, ?}{25} = \frac{100}{25}$$

$$? = 4$$

The amount of BA stock solution you will need is 4.0 ml.

Checklist

When mixing stock solutions or media, a running checklist (Table 6-2) should be followed to ensure that the right chemicals and amounts are used. If the items are checked off the moment they are added to the mixture, there should be no problem if the technician is interrupted.

Table 6-2. Checklist for *Begonia rex* **Medium, Stages I and II**
(Murashige and Skoog 1962; Mikkelsen and Sink 1978)

Chemicals	Stock per liter of medium	7/21/96	8/18/96	9/22/96
Sucrose	30 g	✓	✓	✓
Nitrate stock	10 ml	✓	✓	✓
Sulfate stock	10 ml	✓	✓	✓
Halide stock	10 ml	✓	✓	✓
Phosphate stock	10 ml	✓	✓	✓
Iron stock	10 ml	✓	✓	✓
Inositol/thiamine stock	10 ml	✓	✓	✓
Nicotinic acid/pyridoxine stock	5 ml	✓	✓	✓
BA	4 ml	✓	✓	✓
NAA	1 ml	✓	✓	✓
Glycine (10 mg in 10 ml distilled water)	2 ml	✓	✓	✓
Adjust pH to 5.5		✓	✓	✓
Agar	8 g	✓	✓	✓

Equipment and supplies

1 2-liter Erlenmeyer flask
Distilled water
Hot plate/stirrer and stir bar
Sugar
2 balances
Weighing papers
1 spatula
Stock solutions (the 9 listed above)
1 10-ml graduated cylinder

2 10-ml pipets (in 10ths)
1 1-ml pipet (in 10ths)
pH meter
Medicine dropper
Wash bottle containing distilled water
1 *M* NaOH (for pH adjustment)
1 *M* HCl (for pH adjustment)
Agar
1 pitcher or other dispenser
65 test tubes, with caps
Metal test tube racks
Autoclave or pressure cooker

Procedure for making 1 liter of MS-based *Begonia* medium

1. Pour 600 ml of distilled water into a 2-liter flask. (Always use a container with ample room for the solution to boil. For example, use a 2-liter flask for 1 liter of medium, a 4-liter flask for 3 liters of medium, and so on.)
2. Slide the magnetic stir bar into the flask.
3. Place the flask on the hot plate/stirrer and turn on the stirrer.
4. (Refer to the checklist in Table 6-2.)
   a. Weigh 30 g of sugar on a balance and add to the flask.
   b. For those stock solutions for which 10 ml is required—nitrate, sulfate, halide, phosphate, iron, and inositol/thiamine solutions—pour 10 ml of each solution into a 10-ml graduated cylinder and add to the flask. (It is *very* important to clean and rinse the graduated cylinder well between measuring solutions.) Measure amounts to the bottom of the *meniscus* (the curved surface of a liquid; see Figure 6-1).
   c. Add the required amount of the remaining stock solutions using a clean 10- or 5-ml pipet for each stock solution. Note: whenever you use a pipet, it is wise to use a bulb pipetter, or pipet filler; if a pipet is used like a straw, too often the result is a mouthful of chemicals.
5. Turn off the stirrer and add distilled water to the flask up to the 1-liter mark.
6. Turn on the stirrer and adjust the pH of the medium:
   a. Calibrate the pH meter according to the manufacturer's instructions.
   b. Using a wash bottle with distilled water, rinse off the pH-meter probe.
   c. Lower the probe into the medium as it is mixing.
   d. Observe the reading on the pH meter. If the pH is below pH 5.5 (too acid), use a dropper to slowly add, one drop at a time and allowing the medium to mix after each drop, 1 *M* NaOH until the pH meter reads pH 5.5. If the pH is above pH 5.5 (too alkaline), then slowly add 1 *M* HCl, one drop at a time, until the pH meter reads pH 5.5.
7. Weigh and add 8 g of agar.
8. Turn on the hot plate/stirrer heat. Continue to heat and stir until the medium boils vigorously, but do not allow it to boil over.
9. Turn off the hot plate/stirrer and remove the flask.
10. When the medium stops boiling, add distilled water to the 1-liter mark.
11. Dispense the medium at 10 to 15 ml per test tube using a pitcher or automatic pipetter. Avoid spilling any medium on the rims of the test tubes because the

cotton filter in the cap will stick to the rim of the test tube and come out when transferring. A trace of vaseline placed on the underside of the pitcher lip will help prevent drip.

12. Cap the test tubes.
13. Place the test tubes in metal test tube racks in the autoclave. Alternatively, place the test tubes in the wire basket of the pressure cooker; if there are too few test tubes to stand up properly in the basket, fill the gap with an empty beaker.
14. Process the test tubes for 15 minutes at 15 pounds pressure according to the instructions in the manufacturer's guide.
15. After processing, and when the cooker or autoclave pressure level is back to zero, remove the racks or basket of test tubes and place on an angle so that the medium will solidify on a slant.
16. When the agar has cooled, label and store the sterilized test tubes of medium in a cool, clean place. Store the test tubes in the box in which they were shipped (on its side with thin cardboard sheets between the layers) or in test tube racks (which can be an expensive method of storage when hundreds of test tubes are involved).

LIQUID MEDIA

A liquid medium may be prescribed in some formulas for one or more stages of growth. Liquid media are faster to make than agar media because there is no agar that needs to be dissolved before the medium can be evenly dispensed. Changing to a liquid medium may also help solve some of the problems encountered on gelled media, such as *bleeding* (the exudation of substances—phenolic exudates—which discolor the agar), brown leaves, or poor growth.

Usually liquid cultures are aerated by agitation on a rotator, shaker, or rocker. Agitated liquid media will not allow waste products to build up adjacent to the culture, as can occur in gelled media. Only 5 to 10 ml of liquid medium should be used in a 25×100 mm test tube.

Some cultures in liquid do not seem to require aeration. Sometimes when cultures are started from explants that are 1-in (2.5-cm) long cuttings, or longer, the base of the explant may be placed in 3 to 5 ml of liquid medium in a test tube and no agitation is necessary since the top of the explant is exposed to the air.

Bridges and rafts

Especially for smaller cultures (0.5 to 5 mm), an alternative to agitation is to insert a bridge in the test tube to support the culture, thus allowing it to have air, and convey the liquid to it (Figure 6-2). To make a bridge, cut 3×0.75 in (7.5×1.5 cm) paper strips of filter paper or paper towel. Insert the strips into the test tubes with both ends down so that the middle of the strip will be above the liquid and can hold the culture. This will, of course, be done prior to sterilizing the medium.

Another type of bridge can be made in the transfer hood. Place 3-in (7.5-cm) squares of paper towel or filter paper in a beaker, cover with aluminum foil, and sterilize for 45 minutes at 15 pounds pressure in a pressure cooker or autoclave. Have sterile test tubes ready containing 5 ml of sterilized liquid medium. In the hood, firmly grasp a piece of the toweling using sterile forceps, holding it between the forceps with the tips of the forceps at the center of the paper. With a twist of the wrist, place the paper into the test tube. If it is not grasped firmly, the forceps will simply make a hole in the paper and slide through. When the paper is partially in the liquid in the bottom of the test tube, fold over the corners to make a platform on which to place the

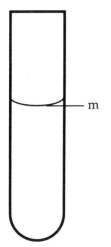

Figure 6-1. Liquid in a test tube showing the bottom of the meniscus (m), which is due to surface tension

m

Figure 6-2. Test tube showing filter-paper bridge in liquid medium.

culture. The liquid medium will wet the paper without submerging the culture. A variety of rafts, floats, bridges, or other supports are available commercially, but most are expensive and/or cumbersome.

Certain general guidelines for media preparation and explant pretreatment will increase the chances of starting explants successfully. If the prescribed formulas for culturing the plant at hand prove ineffective, try various standard media (both liquid and agar), starting with those that have been used successfully for related plants in the same family or genus. A useful approach for herbaceous perennials is to start them in half-strength MS salts without hormones; for woody plants, try WPM without hormones.

By now you should have a feeling for how to mix media. If you have followed the above protocol, you are ready to find out how your plants will do in your media. There are about 900 species of *Begonia* to choose from. Hopefully, most of those that you try will respond favorably. Let curiosity be your guide; probably some unrelated plants will respond equally well to the same formulas. Read on to learn about preparing the explants for the media you have ready.

Chapter 7: Explants and Their Preparation

An explant is a piece of a plant from which a culture is started. Theoretically, a single explant can produce an infinite number of plants. This is probably the most illuminating and the most astonishing statement one can make about plant tissue culture. Indeed, if all goes well, one explant will produce thousands of plants. Of course, it is good to have more than one explant in order to achieve the desired number of plants in less time, and also because typically some number of plants will be lost as a result of contaminants, disinfectants, or other unknown reasons. Normally, however, only a few explants are necessary, and one stock plant is usually sufficient to supply all the starts required.

Explants range in size from a microscopic 10th of a millimeter to stem pieces of several centimeters in length. Explants can be meristems, shoot tips, macerated stem pieces, nodes, buds, flowers, peduncle (flower stalk) pieces, anthers, petals, pieces of leaf or petiole, seeds, nucellus (the central part of an ovule) tissue, embryos, seedlings, hypocotyls, bulblets, bulb scales, cormels, radicles, stolons (runners), rhizome tips, root pieces, or, though rarely, single cells or protoplasts. (Figures 2-3, 7-1, and 7-2 provide illustrations of many of these plant parts.) Quite often the misinformed lay person will ask, "You start with a single cell, don't you?" To which I reply, "Heavens, no!"

While no precise instructions can guarantee successful explants, many protocols have been universally successful, and general guidelines offer a plethora of alternatives. One may well ask, When is the best time to obtain an explant? A stem tip may perform best if taken a little earlier in seasonal development than one would take a normal cutting. However, the odds of success are greatest if a few explants (10 to 20) are started at weekly or monthly intervals, instead of gambling many valuable explants on only one set of conditions. For *Begonia rex*, the example for tissue culture procedures used in this book, if the plant to be used is growing well indoors, it should not need any special treatment prior to obtaining an explant.

PLANT PATHOGENS

Only disease-free plants should be tissue cultured. Sometimes a plant that appears healthy is harboring a disease. If time permits, and depending on the particular plant and the customer, the plant should be tested for microbial organisms known to infect that species. Testing by a government or independent laboratory is usually more acceptable to a customer (due to impartiality), and government testing is required for shipping certain species across some state or national borders.

Germ plasm repositories (government locations for collecting and preserving live-type species) are good sources of plants that are free from pathogens. At these laboratories technicians heat treat plants, grow new plants from meristems, grow material in screen houses, test for diseases, and provide clean stock for commercial users. Increasingly, plant propagators use pathogen detection kits, which are commercially available, to identify fungal, bacterial, and viral pathogens. Another common method of testing for disease is to graft shoots on indicator plants that are sensitive to specific diseases and will show symptoms in a matter of hours or days. Specific information on identifying and treating for contaminants is given in Chapter 11.

If it is discovered by observation or testing that a valuable plant is infected with a viral disease, it is sometimes possible to rid the plant of the disease by heat treatment, followed by meri-

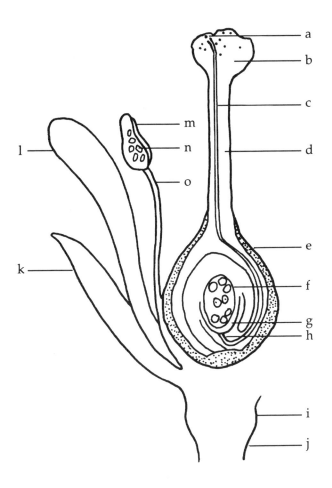

Figure 7-1. Flower parts: a. pollen grain; b. stigma; c. pollen tube pathway; d. style; e. ovary; f. ovule; g. nucellus; h. micropyle; i. disc; j. receptacle; k. sepal; l. petal; m. anther; n. pollen sac; o. filament.

stem culture. Heat treatment is commonly applied under good light (500 to 1000 f.c.), with a 16-hour photoperiod. The length of time and the amount of heat applied vary with the tolerance of the plant, the part of the plant used, and the type of virus. For example, *Pelargonium* plants have a low heat tolerance and will seriously etiolate (blanch) after about 3 weeks of heat treatment, whereas carnations *(Dianthus)* have a high tolerance and can survive heat treatment for several months. Only healthy appearing plants with adequate energy reserves should be heat treated, because exposure to temperatures above optimal-growth temperatures is quite detrimental.

OBTAINING EXPLANTS

Before taking explants, stock plants should be moved to a clean greenhouse, screen house, or other shelter away from dust and disease. There they can be observed and given special care for a time, usually about 2 weeks to 6 months. In controlled conditions the plants will become as healthy as possible, and they will not be stressed, as they might in a field or garden. Wash the plants with clean water, allow the foliage to dry, and thereafter feed and water them only at the base. The new shoots that appear will provide relatively clean explants and, in addition, may provide some increased juvenility.

Just as juvenility is a factor in selecting material for cuttings (Hartmann et al. 1981), so it is in selecting explants for tissue culture. In general, the more juvenile the explant material, the

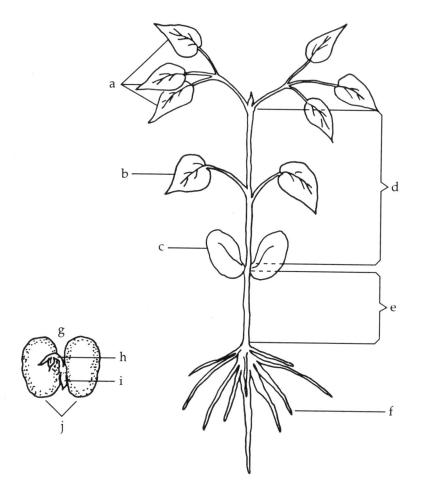

Figure 7-2. Seedling parts: a. true leaf; b. first leaf; c. cotyledon; d. epicotyl; e. hypocotyl; f. root. Bean seed (split open): g. embryo; h. plumule; i. radicle; j. cotyledons.

greater the likelihood of success. With this in mind, any means by which more juvenile material can be taken is preferred. The juvenile leaves of *Eucalyptus*, *Sequoia*, and *Vitis* are visibly different in form from their mature leaves. As mature plants these genera sometimes produce adventitious shoots near their bases. These shoots have juvenile leaves, a fact that lends credence to the theory that the bases of plants may be more juvenile in their lower reaches than higher up. Sometimes plants can be repeatedly pruned back (hedged or sheared) to encourage regeneration of juvenility in new growth, which then can be taken as explants. Grafting mature shoots on juvenile rootstock can also foster some increased juvenility. Growth from rooted cuttings of mature plants is not considered juvenile material.

When shoot tips, stems, buds, or flowers are taken for explants, they should be cut longer than the final size you plan to use. The final length may be 1 in (2.5 cm) or more, or it may be the microscopic apical meristem. In most instances, about one inch is a convenient size to process. Any final trimming and treatment should be done in the sterile environment of the transfer hood.

Occasionally leaves or parts of leaves are used to start tissue cultures. It is common knowledge that African violets *(Saintpaulia)* and a number of other plants can be propagated from leaves. Some conifer needles in culture will produce plantlets around their edges. Other leaves, after cleaning, can be drilled with a cork borer and the discs placed in culture. Leaf tissue is also a common source of protoplasts, the culture of which is discussed in Chapter 13.

FURTHER CONSIDERATIONS

- The smaller the explant, the less contamination to remove; but the larger the explant, the more tissue there is to help establish it in culture.
- Use new shoots whenever possible because they are cleaner than old wood.
- Plants under cover or shoots from forced, dormant branches are usually cleaner than field plants.
- Explants from younger plants often respond more quickly than those taken from older plants.
- Plant material at the base of a plant is deemed younger than growth higher up. Plant material at the base of a plant also can be less clean, however, due to proximity to the ground.
- If plastic bags are fastened over actively growing shoot tips on stock plants, the new growth will be cleaner for cutting explants days or weeks later; be sure to provide shade to prevent sunburn.
- Explants may do better if taken in the morning, rather than later in the day.
- Cut explants with cutters that have been sterilized by dipping in 1/10 bleach.
- Upon cutting explant material from the source plant, place the explant in a plastic bag containing a moist (not dripping wet) paper towel. The explant should remain in the bag until you are ready to work on it in the laboratory.
- Keep explant material refrigerated until you are ready to process it, but for best results process explants as soon as possible.
- If using tips or pieces of roots or rhizomes as explants, they will be easier to clean if the plant has been growing in an artificial medium, such as washed sand, pumice, or perlite. Wash such cuttings before placing them in a plastic bag.
- Leaf petiole sections 5 mm in size are recommended for explants, but a section of petiole together with a centimeter of leaf base or a node or shoot tip will also do well.
- Petiole pieces of 2 to 3 cm or so are easy to manage; cutting to 5-mm pieces should not be done until final processing in the hood.

EXPLANT CLEANING AND TREATMENT

After explants are detached from the source plant, they must be submitted to cleaning treatments. There are no set rules for cleaning explants because the amount and kind of cleaning depends on how clean or dirty the plant material is to begin with. The process selected also depends on how easily the explant is damaged by a cleaning agent. Disinfectants will not eliminate endogenous (internal) contaminants, and exogenous (external) contaminants will persist if the disinfectant solutions used in treatment are too weak, especially those contaminants located in plant crevices.

The most common cleaning agent used to disinfect explants is bleach in various concentrations (pure household bleach is 5% sodium hypochlorite), used alone or in sequence with other disinfectants. Commonly used concentrations of household bleach solutions are 5%, 10%, and 20% concentrations. For 100 ml of bleach solution, mix 5 ml of bleach with 95 ml of sterile water to make a 5% concentration; 10 ml bleach with 90 ml sterile water for 10% concentration; and 20 ml bleach with 80 ml sterile water for 20%. These concentrations are often referred to as 1/20, 1/10, and 1/5, respectively. Thus 1/10, for example, means 1 part bleach *in* 10 parts of solution. If directions specify 0.05% sodium hypochlorite, that would indicate a 10% solution of household bleach.

If the explant material is very dirty, one helpful method of pretreatment is to place the explant in running water overnight. Another approach is to agitate the material in 1/100 (1%) bleach, plus 2 drops of Tween 20 per 100 ml of solution, for one or more hours. Tween 20 is a wetting agent (surfactant) that reduces the surface tension of the material being cleaned, thus allowing the bleach to be more effective. Liquid dishwashing detergent can have the same effect but should be used with care because some kinds of detergents may damage the explants.

Other disinfectants commonly used to clean explants include 70% ethyl alcohol or isopropyl alcohol; 5 to 10% calcium hypochlorite; or 3% hydrogen peroxide (H_2O_2). Isopropyl alcohol, commonly recommended in the literature, is readily available as rubbing alcohol; ethyl alcohol is more difficult to come by and is more expensive. If you have enough explants, try treating several small groups with various concentrations and various kinds of cleaning agents, and vary the time of immersion in the disinfectant. *Use each disinfectant separately; never mix any 2 disinfectants in the same container, because dangerous chemical reactions will occur.*

The following is an example of an explant cleaning treatment that is used for 10 1-in (2.5-cm) strawberry *(Fragaria)* runner tips. It is performed in a series of beakers that have been autoclaved and rinsed in 10% bleach. The recommendations for time and bleach concentration may vary according to the tolerance of the explant and the effectiveness of the treatment.

Equipment and supplies
> 5 100- to 250-ml beakers
> 1% bleach
> 10% bleach
> Dropper bottle containing Tween 20
> Magnetic stir plate and stir bar
> Explant material
> Distilled water
> 70% isopropyl alcohol
> Forceps
> Cutters or scalpel

Procedure
1. Pour 100 ml of 1% bleach and 2 drops of Tween 20 into a sterile 150-ml beaker with the magnetic stir bar and place on the stir plate.
2. Add the explants to the beaker and stir for 10 minutes. (As an alternative to the stir plate, the explants and the cleaning materials can be shaken by hand in a tightly closed jar, or they can be agitated on a shaker.) Pour off the bleach solution, and place the explants in a clean beaker.
3. Rinse the explants with distilled water. Pour off the water, and place explants in another clean beaker.
4. Rinse the explants in 70% isopropyl alcohol (or 3% hydrogen peroxide) for 5 seconds. Again, pour out the alcohol and place the explants in a clean beaker.
5. Rinse the explants in distilled water, then place them in a clean beaker.
6. Pour 100 ml of 10% bleach and 2 drops of Tween 20 into a sterile 150-ml beaker with the magnetic stir bar and place on the stir plate.

7. Add the explants to the beaker and stir for 15 minutes. (Some tissue culturists use a vacuum pump [at 25 mm of mercury] to help the disinfectant penetrate into the crevices of the explant. Others have found that an ultrasonic cleaner is useful in cleaning explants. Still others simply shake by hand a tightly closed jar containing the explants and solution.)

8. With the explants still in the beaker of 10% bleach solution, move explants to the transfer chamber.

If the external contaminants are not killed by the foregoing treatment, microbial growth, usually fungi, will be evident in the test tubes within a few days. If such growth is observed, then it will be necessary to treat new explants incorporating one or both of the following alternatives:

- Stir the material in the 10% bleach for 5 to 15 minutes longer.
- Extend the alcohol dip to a minute or longer.

If any of the preceding treatments kill the explants, try weaker bleach solutions for the same or longer time periods. To test the tolerance of a set of explants to 10% bleach, remove a few of the explants at 5, 10, and 15 minute intervals, instead of running them all for 15 minutes. Sometimes a weaker bleach solution for a longer period of time is more effective. For example, try agitating explants in 5% bleach for one hour, or in 1 to 3% bleach overnight.

The cleaning of explants should be done in the media preparation room up until the final stages of explant preparation, both for convenience sake and to avoid the clutter of a stirrer/shaker, dirty rinse containers, disinfectant containers, and other paraphernalia that might contaminate the transfer hood. Transport the cleaned explants to the transfer hood, still in bleach in a covered container, and apply the final rinses in the hood using sterile technique. This final stage and sterile technique are presented in the next chapter.

Chapter 8: The Transfer Room

Even though you may be more used to operating in gardens and greenhouses than in test tubes and jars, you will soon be at ease working with plants in miniature. You will feel the same fascination and wonder, the same concerns, cares, joys, frustrations, and curiosity that plants inspire in any setting. You will learn to detect subtle changes in growth or responses to different media, and you will recognize symptoms of stress and disease that may not now be familiar to you. Sometimes, unfortunately, you will observe pinpoints of contaminating molds, or wisps of bacteria in a medium. You will be reminded that the air is full of contaminants, that much greater care is required than in your greenhouse, but also that contamination will decrease as your technique and knowledge increase. Because you are familiar with the whims of plants, you know that patience and persistence are crucial for any successful crop.

Thousands of transfers are made in the transfer hood using sterile technique. The operations will go very slowly at first, but with practice you may eventually transfer 1000 or more cultures a day, depending on the material. You will have to make many decisions based on careful observation of the cultures during transfer. Watch for plants that should be culled. Make note of what media and techniques work best, and of what changes might be made.

STERILE TECHNIQUE

As we have emphasized, cleanliness is of primary importance in the transfer room, especially in the hood, where cultures are momentarily outside of their sterile protective containers—a time when they have the greatest danger of becoming contaminated. It is not easy to imagine and deal with the unseen. Microscopic organisms or contaminated particulates are literally everywhere, except on that which has been sterilized and subsequently protected. One has to assume that clothing, skin, counter tops, instruments, and the air are teeming with mold spores, bacteria, and a host of other invisible enemies ready to invade clean cultures. Just think where you would be if you were a bacterium.

Air is sterilized by filtration. The air gently streaming through the transfer hood (see Figures 4-12 and 4-13) is filtered so as to allow the technician to open culture containers and make sterile transfers with reasonable assurance that the cultures will not be contaminated during the operations. Except for the HEPA filter, the hood should be wiped down daily with a disinfectant such as 10% bleach, 70% alcohol, or Lysol or some other household disinfectant. HEPA filters are fairly fragile and any contact with them should be avoided; take care to not push containers, instruments, or other objects into a filter. In addition, when arranging equipment in the hood, avoid any obstruction to the laminar air flow because it will change the air flow pattern and so invite contamination. Do not place any possibly contaminated items such as beakers or racks of test tubes between the HEPA filter and exposed cultures.

Be sure to review the section in Chapter 4 on sterilizing equipment for transfer tools. The more commonly used methods include alcohol, bleach, Bacti-Cinerators, and glass bead sterilizers. Each method of sterilizing has its drawbacks. The alcohol flame poses a fire hazard. Bleach must be mixed fresh every 4 hours because the chlorine fumes given off from open containers of bleach can be harmful or irritating. Bleach is also a little messy to work with. When using bleach, wear gloves because it can burn your skin; it will also stain or eat holes in some fabrics. The electric sterilizers heat the instruments so hot it takes them a long time to cool; plus they are more expensive. Deciding which is the best sterilization technique is a matter of individual preference. Whichever method you choose, it is usually best to have more than one set of instruments so that one set can be sterilized, cooled, or dried while the other is being used. Instruments should be stainless steel, otherwise they will corrode in any of the sterilizing methods. The procedures defined here employ bleach solutions for sterilizing instruments. Using bleach is a good method for the beginner, especially the beginner who wishes to spend as little capital as possible to start with. The same general principles apply to other sterilization methods.

The following practices should be strictly observed at all times to maintain safe, clean conditions in the laboratory:

- No eating, drinking, smoking, or storing food in the laboratory.
- Wear a protective lab coat or apron when working with cultures. Long, full sleeves should be avoided because they collect dust.
- Wear sterilized disposable medical gloves or plastic garden gloves.
- Tie-back and cover long hair. Loose hair is a potential contaminant for cultures and also risks catching fire from any open flames.
- Sit straight and keep your head out of the transfer hood. Work at arm's length, as far back in the hood as is practical. Avoid wide sweeping arm movements over the work area; if a hand or arm reaches from one side of the hood to the other, contaminants can fall from sleeves or arms. Keep your right hand on the right-hand side and your left hand on the left-hand side of the transfer hood work area. If you need to pass an object from one side of the hood to the other, use great care, keeping elbows as close to the body as possible; or simply pass the object from one hand to the other. The hand that holds a test tube is contaminated by it and can in turn contaminate forceps, which if not sterilized, can then contaminate the inside of a test tube during a transfer.
- Set only necessary equipment in the transfer hood.
- Sterilize in an autoclave or pressure cooker any jars and petri dishes containing contaminated cultures before emptying.
- Disinfect work surfaces before and after working in the hood.
- Keep transfer room doors closed to avoid unnecessary drafts.

It is best to start cultures singly in test tubes, instead of in jars, because in this way if one start becomes contaminated it will not contaminate others, as it would if surrounded by others in a jar. If a hundred or so test tubes of a particular plant are being propagated, and the loss of a few in one jar would not be a significant loss, then it is time to move them into jars. Also remember that, because it is difficult to distinguish one culture from another, to ensure accurate labeling it is helpful to devise a scheme and to have a second person check your labeling. To further avoid mislabeling, it is important to work on only one clone in the hood at any one time.

Having considered some broad matters of sterile technique, we can now address the specific activities in the transfer room. In following the procedures, be sure to employ the general practices for maintaining a clean and safe environment listed above. The technical operations involved in final explant preparation and in dividing and transferring plant material that is already started and multiplying demonstrate the set pattern for all operations in the laminar air flow hood. Other procedural modifications, such as using another system for sterilizing transfer tools, will require only slight adjustments, which will be self-evident to you, the technician.

Equipment and supplies

    10% bleach in sterile plastic dish

    1% bleach in sterile plastic dish

    2 100-ml sterile beakers, each containing about 75 ml of sterile distilled water

    1 100-ml sterile beaker containing 1% bleach solution

    Plastic gloves

    Forceps

    Stainless steel knife with disposable blade

    Instrument holder

    Pretreated explants in beakers of 10% bleach

    Sterile commercial-grade paper towels

    Sterile test tubes of MS-based *Begonia* medium

    Test tube racks for 10 and 40 tubes

    Wastebasket, placed on the floor beside the technician

    Labeler

    Planter trays or Speedling trays (or test tube racks)

When preparing the bleach, disinfect 2 plastic dishes with alcohol and rinse well with distilled water before pouring in the bleach solutions. Alternative methods for sterilizing transfer tools—such as glass bead sterilizers or Bacti-Cinerators—can be used in the transfer hood as well. To sterilize the paper towels, wrap about 25 towels in aluminum foil or place in an autoclavable plastic bag and sterilize for 45 minutes at 15 pounds pressure.

STERILE TECHNIQUE FOR EXPLANTS

    The following procedure picks up where step 8 of explant cleaning and treatment in Chapter 7 leaves off. Left-handed people should reverse the hands indicated here.

Procedure

    1. Turn on the transfer hood blower for 10 minutes before transferring (or leave it on all the time).

    2. Rinse gloved hands in 10% bleach.

    3. Wipe or spray inside the hood with 10% bleach, 70% alcohol, or Lysol; do not spray the HEPA filter.

    4. Rinse gloved hands in 10% bleach.

    5. Immerse knife (scalpel) and forceps in 10% bleach for at least one minute and rinse in 1% bleach. Drain and place in instrument holder.

    6. Using forceps, rinse the treated explants in 1% bleach, then in 2 sterile distilled water rinses. Leave the explants in the sterile water rinse while performing steps 7 and 8.

    7. Using forceps and knife, lay a sterile paper towel on the counter as far back in the hood as is practical to work.

    8. Return forceps and knife to 10% bleach for one minute, then rinse in 1% bleach.

    9. Using forceps, place an explant from the sterile water onto the paper towel.

    10. Pass the forceps to the left hand and pick up the knife with the right hand.

    11. Using forceps and knife, trim the explant appropriately. (See Section II for guidelines on trimming explants for particular plants.) For shoot tips, you can use from 1 mm to several centimeters; 2 to 3 cm (about 1 in) is usually a good

length. Stem or petiole sections should more often be just a few millimeters. Whole young buds will naturally vary in size.

12. Return forceps and knife to 10% bleach for one minute, then rinse in 1% bleach.

13. With the left hand, grasp a test tube containing medium for the explant. Hold the test tube near its base (see Figure 8-1). To best prevent any microbes from falling into the test tube, hold it at about a 50° angle and facing right—parallel with the front of the hood so that it is facing neither the back filter nor you.

14. Take forceps in the right hand. While still holding the test tube and the forceps, grasp the test tube cap with the right-hand little finger, push and twist the cap slightly (to help keep the cotton in the cap), and remove the cap from the test tube.

15. Still holding the test tube, forceps, and cap, use the forceps to remove an explant from the sterile paper towel and place the explant firmly on the agar in the test tube (Figure 8-1).

16. Still holding the forceps, replace the cap on the test tube and place the tube in the test tube rack.

17. Return forceps to 10% bleach.

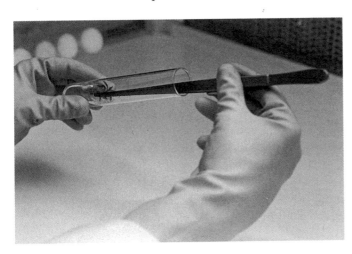

Figure 8-1. Technique for inserting a propagule (transfer) into a test tube. The test tube cap is hidden, held by the right-hand little finger.

STERILE TECHNIQUE FOR TRANSFER

The sterile technique for transferring cultures is similar to that of the final stages of explant treatment just described, yet it is different enough to require redefining some of the steps.

Procedure

1. Place the test tubes containing cultures ready for transfer on a counter or cart outside of the transfer hood.

2. Have fresh sterile test tubes containing agar medium ready to receive the transfers.

3. Steps 1–5 above.

4. Steps 7 and 8 above.

5. With the left hand, grasp a test tube containing the culture to be transferred. Check the label to confirm that the test tube contains the correct culture.

6. As described in steps 13 and 14 above, hold the test tube near its base and at

about a 50° angle. While holding sterile forceps in the right hand, twist off the test tube cap with the right-hand little finger. Remove the cap and place it on the work bench (to be removed later), or drop it into a temporary container on the floor.

7. Using sterile forceps, remove the culture from the test tube and place it on the paper towel in the transfer hood.

8. Place the test tube in a 40-tube test tube rack, which can go directly into the dishwasher.

9. Pass the forceps to the left hand and pick up the sterile knife with the right hand. Cut, trim, and divide the culture as necessary (see Section II). Trim away any brown or dead material.

10. Return knife to 10% bleach.

11. Pass forceps to the right hand.

12. Using forceps, place the culture in a test tube with the appropriate medium, following the procedures described in steps 13–15 above. Some shoot transfers should be laid on the surface of the agar to induce bud break if there are lateral nodes. Most shoot tips should be inserted into the medium with their bases just deep enough in the agar (about one centimeter) to allow the shoot to be held upright. Transfers that are going into a rooting medium should be held upright.

13. With forceps still in the right hand, replace the cap on the tube and place the test tube in a rack ready for labeling.

14. Return forceps to 10% bleach.

15. After doing about 6 test tubes, place the towel with the disposable material on it into the wastebasket, and get a fresh sterile paper towel, again using sterile technique to lay the towel down in the transfer hood.

16. When you have finished transferring all the tubes of one variety, label each test tube with a marking pen, grocery-store labeler, or hand stamp. Code the date, culture name, and your name.

17. Place the test tubes in planter trays (or Speedling trays) or keep in test tube racks, and move the transferred cultures to the shelves in the culture growing room.

18. Record the date, how many test tubes in, how many tubes out, multiplication rate, medium used, when next transfer is due, your initials, and any other desired information.

OTHER TISSUE CULTURE TECHNIQUES

MERISTEM CULTURE

The term "meristem culture" is often used incorrectly to indicate micropropagation. Literally and correctly, however, it refers to the utilization of meristematic tissue for explant material, usually the microscopic, meristematic, apical dome and the accompanying pair of leaf primordia. Meristemming is frequently used for establishing virus-free clones, the idea being that meristematic tissue grows faster than plant viruses can infect these new cells; therefore, the smaller (i.e., younger) the meristem explant, the greater the probability that it is virus free. Some government laboratories use heat treatment, take meristems, and test for viruses before issuing virus-indexed plants (see Chapter 11).

Strawberries (*Fragaria*) are easy to meristem compared with most plants, but the process for treating and extracting strawberry meristems is quite typical of the operation and therefore makes a good plant to use for learning the procedure. For these reasons it is described here (Figure 8-2).

To meristem a strawberry *(Fragaria)* plant, select a young runner before the bud leaves have expanded. Cut about 2 in (5 cm) of the young runner tip. Because the meristems are within protective buds, initially they are clean; careful treatment should be undertaken to ensure that no contaminants are introduced.

Equipment and supplies

Pretreated explant material
10% bleach in sterile beaker
70% isopropyl alcohol
Dissecting microscope
Distilled water in beakers
Sterile petri dish
6-in (15-cm) stainless steel knife with disposable blade
Small (watchmaker's) forceps or tweezers
Sterile paper towels
Test tubes containing sterile agar medium

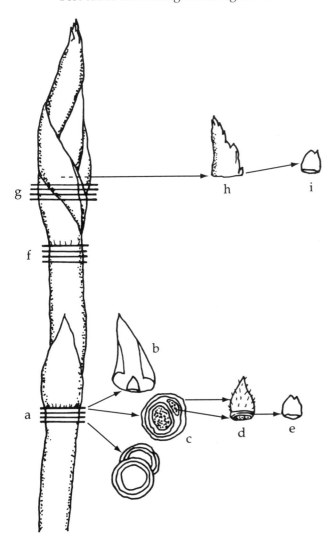

Figure 8-2. Strawberry *(Fragaria)* runner tip: a. first series of cuts—start at the bottom line; b. bract with attached bud; c. cross section of stem, showing base of bud; d. removed bud (magnified); e. meristem showing 2 primordial leaves; f. second set of cut cross sections; g. third set of cross sections; h. bud with attached meristem; i. meristem showing 2 primordial leaves

Procedure

1. Stir the explant in 10% bleach for about 10 minutes.
2. Wipe down the dissecting microscope with 70% isopropyl alcohol before placing it in the transfer hood. Do not touch the microscope lenses.
3. In the hood, agitate the explants in 2 rinses of sterile distilled water.
4. In the hood, place a sterile petri dish under the objective of the stereo dissecting microscope.
5. Using sterile technique, place a piece of sterile paper towel in the petri dish and moisten with sterile water.
6. Using sterile technique, place the runner tip on the paper towel in the petri dish under the microscope objective. Have ready a 6-in (15-cm) knife with disposable blade and watchmaker's forceps (or tweezers).
7. Search for 2 or 3 buds on the runner tip. At some distance from the tip you may notice a bulge in the stem, with or without an accompanying bract.
8. Using sterile technique, with the knife slice across the stem just below the bulge. Make another horizontal cut at the point of attachment of the outer epidermal layer.
9. Make a vertical slit through the loosened outer layer and tear it away with the forceps. Examine the base of the stem piece to find the bud. If very young, the bud may be difficult to detect, lying flat and close to the stem. If you cannot locate the bud at the base, then examine the base of the sheath you removed.
10. When you locate the bud (some runners may not have them in this area), dissect it by slicing away very thin slices from the base up to the meristematic dome. The dome and 1 or 2 pairs of leaf primordia will appear clearer than surrounding tissues.
11. Touch the knife tip to this bit of tissue. Moisture will hold it to the knife tip until you place it in the test tube and release it by lightly cutting the blade tip into the agar.
12. Cap the test tube, then examine it while holding it up to the light. You will barely be able to see the microscopic explant.
13. To locate the next bud, continue to slice away the stem upward toward the tip until you find another bulge.
14. Again, slice through the stem at the point of attachment of the sheathing, or outer layer. Find and dissect the bud as described in steps 10 and 11.
15. The third bud is the apical bud and will be more difficult to find because of the older leaves surrounding it. Continue to slice upward on the stem. Often a slice will reveal the base of a bud, or the dome may actually fall out.

Knowing precisely what to look for and where will come quickly with experience and will be very rewarding.

MACERATED TISSUE CULTURE

Macerated tissue is one form of explant material that is used for tissue culture, particularly for growing callus cultures. Individual stem tips or meristems can be mashed with a scalpel, knife, or spatula. Sometimes cleaned explants can be homogenized in a sterilized blender. This method has worked for Boston fern *(Nephrolepis)*. Any macerating of explants should be done in the sterile environment of the transfer hood.

NURSE CULTURE

Nurse culture is a technique that was developed to obtain clones of cells arising from a single cell. It is a technique worthy of consideration for stimulating growth of difficult-to-grow cultures. For example, although many orchids have been successfully propagated by tissue culture, many others are still awaiting development of a suitable method for *in vitro* propagation.

In this technique, a filter paper (a semi-permeable barrier) is used to separate the nurse culture (usually callus tissue, either in a liquid or agar medium) from the single cell or cells. The cells receive through the filter paper nutrients and growth factors released from the callus and from the medium (Horsch et al. 1980). The growth factors provided by the nurse tissue induce growth and division of the cell or cells on the filter paper. This technique puts to practical use the fact that a growth medium conditioned by one tissue can promote growth of a low-density culture of a different tissue. Evidence such as this suggests that metabolites critical to initiation of cell division may be the same in cells of different species.

MICROGRAFTING

Micrografting is a microscopic version of normal grafting. Although not widely practiced, in principle meristems can be grafted onto virus-free rootstock to eradicate viruses from the scion plant. The practice also has potential for viral assays and as an alternative to conventional greenhouse grafting. As with normal grafting, the cambium of scion and rootstock should match. The scion can either be placed on top of the rootstock or inserted into a T epidermal incision. Rootstock can be grown in the greenhouse or tissue cultured, or it can be grown from *zygotic embryos* (embryos from seed). *Vitis* meristems have been successfully grafted on *Vitis* zygotic embryos (Zimmerman 1991). White pine *(Pinus)* meristems have also been successfully grafted on white pine zygotic embryos. In the case of the white pines, it was determined that the preferred point of graft for both juvenile and mature meristems was on the embryo hypocotyl midway between the radical and the base of the cotyledons (Goldfarb and McGill 1992).

CULTURE MANIPULATION

Once an explant culture is started, whether from a meristem, a shoot tip, a leaf stem, callus, or whatever, the question is what to do next. It is important to remember that you should never open a test tube or other container with a clean culture growing in it, other than in the transfer hood using sterile technique.

Depending on the plant and on the stage of the culture, there are numerous ways to divide and transfer growing cultures. The early stage of a strawberry *(Fragaria)* meristem culture, for example, is a little green mound that is simply cut in 2. A new culture of a *Rhododendron* shoot tip, on the other hand, requires that new shoots be cut off and transferred and the old shoot transferred to allow other buds to elongate (see Figure 8-3). An older *Rhododendron* culture will have top growth that lends itself to microcuttings. The top growth of some lily *(Lilium)* cultures is to be cut off and discarded in order to encourage more basal growth, which is then divided. Frequently, cultures are divided in a manner whereby the transferred piece will have a surface area of about 1 square centimeter and have one or more shoots. Too small a division will take too long to recover and multiply. Too large a division will need to be transferred too soon for economy's sake. Section II describes the best methods for dividing and transferring specific plants.

Do not be in a hurry to divide a new explant. Often, particularly with woody plants, it is best to leave a portion of the original stem, with new growth growing from it, for 2 or 3 transfers, gradually removing the old stem. For example, allow a new lateral shoot on a *Rhododendron* explant to grow to about 1 in (2.5 cm) in length before removing all of the original stem. In the *Rhododendron* shoot tip culture illustrated in Figure 8-3, the top bud of the original explant is removed to help force the lateral buds. When one of the top lateral buds grows out, the explant is cut horizontally (b), on sterile paper towel in the transfer chamber, and the 2 pieces are placed in separate test tubes, as illustrated. When the other top lateral bud grows out on the dissected piece, a vertical cut is made (c) to produce 2 new cultures (e and f). Meanwhile, as the left lateral bud grows out on the base piece, a second horizontal cut is made (d); the excised bud, together with some of the original stem, is placed in a new tube (g). The remaining part of the original cutting (h) still contains a bud, and that may yet grow out. In the next transfer, which is not shown, the top buds will have developed sufficiently to be cut from the original main stem. The growth shown here would take place over several months.

Meristem explants are usually slower to develop than shoot tips. The growth from a strawberry *(Fragaria)* meristem may be readily visible within a week, but it may not be more than 2 to 3 mm after one month. At this size (and/or time) the meristem will benefit from rolling it over. Simply move the meristem explant (using forceps and sterile technique) to a new location on the same agar slant. This serves to place the explant in fresh medium and ensures that it has good contact with the agar. In another 1 to 3 weeks it may become a 6-mm green ball or base. Cut this in 2 and place each piece in a new tube.

Cultures can be considered established and in the multiplication stage (Stage II) as soon as they are successfully divided. Then the question is, How big should the transfers be and how often should they be transferred? One rule of thumb is to transfer about a 0.25-in (or about 8-mm) diameter piece when the material has primarily basal growth (mainly for herbaceous plants); or transfer a 1-in (2.5-cm) long piece when the culture is primarily extended shoots or lateral breaks. Usually when clumps develop you should make one or more vertical cuts through the base for new transfers. In blackberries *(Rubus)*, for example, if the tops are more than 1 in (2.5 cm) long, they can be cut off and transferred in small clumps. If tall enough, 1- or 2-node cuttings can be made. Some bulbs do best if top growth is discarded and only the bases are divided and transferred. Shoots or stem pieces should be laid down on agar so as to provide maximum contact between the culture and the agar. However, if leaves bleed where they rest on the agar, it is better to place only the base of the stem piece in the agar.

The length of time between transfers varies according to the plant variety, culture history, size of transfer, media, and other factors. It is a sign that it is past time to transfer when the medium is discolored, leaves turn brown, growth slows or stops, or test tubes become crowded. Ideally cultures should be transferred just prior to any of these events, but that time may be difficult to determine. Experience is probably the best instructor. Some plants require transfer every 2 weeks, others every 6 weeks. The time cycles will vary not only with the species but from one stage to another. Strawberries *(Fragaria)*, for example, may require transfer every 10 to 14 days after the second transfer, but after 2 or 3 months in culture the best interval may be 4 weeks.

The 4 stages of culture growth—establishment (Stage I), multiplication (Stage II), rooting (Stage III), and acclimatization or hardening off (Stage IV)—are not necessarily distinct. In many instances a different medium does not need to be used for each stage because (1) the multiplication medium serves just as well for the establishment stage; (2) rooting may be done *in vivo* (out of culture); (3) a single medium may be satisfactory for the first 3 stages; or (4) a prerooting medium, perhaps one with no hormones so as to improve height, may be applied in the late-multiplication stage.

It may be surprising to people that cultures generally have no roots throughout Stages I

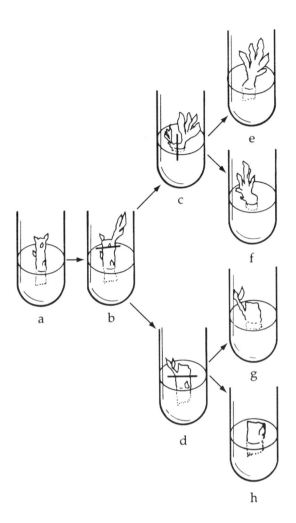

Figure 8-3. A *Rhododendron* shoot tip explant, showing growth and cuts to be made in the first 3 transfers.

and II. The absence of roots appears to help promote shoot growth and multiplication. If roots do appear, the culture will probably stop multiplying and will mature.

CLEAN UP

Tissue culture operations produce a constant flow of used containers that require washing for reuse. For labs that have household dishwashers, the tines on the bottom shelf can be bent down to accommodate racks of test tubes, which are inverted and placed on the shelves for washing. It is not necessary to remove the agar medium from the test tubes before washing; the hot water in the dishwasher will dissolve and wash it away with no problem. To keep the test tubes in the rack, place a hardware cloth (wire mesh) cover over the tubes before turning the rack upside-down and putting it in the dishwasher. Jars can also easily be washed in the dishwasher.

Tube caps with cotton filters seldom need to be washed because they are sterilized with each new batch of medium. When caps appear dirty or wet, remove the cotton, wash the caps, and put in a new piece of nonabsorbent cotton.

Agar should be disposed of promptly so that it will not grow microorganisms. Culture media contain nutrients that are ideal for many contaminants. When cultures become visibly contaminated, the molds, yeasts, or bacteria will quickly overgrow and kill the plantlets in culture. Occasionally media containers will become visibly contaminated before they can be used for transfers. Contaminants form patches of growth in or on agar. Liquid media become cloudy and are no longer clear when they are contaminated. If the agar is opaque (as some agars are) it can be difficult to see contaminants. When contaminants are present they pose a significant threat of spreading to other cultures. It is doubtful that any of the contaminants are human pathogens (most human pathogens thrive in a higher pH), but there is always this risk. The safest means of disposing of these colonies is to place containers that have the contaminants in a pressure cooker or autoclave and sterilize them before opening, emptying, and washing them. However, if microbial contaminants appear in jars of cultures that are about ready for Stage IV acclimatization, the cultures may well be rinsed and planted.

Chapter 9: The Growing Room

The activity in the culture growing room depends on the work in the media preparation and transfer rooms, and the success of culture growth determines the success of the entire business—it is the reason for precise media making and painstaking transferring, and is the basis by which viable plants will sell. Knowledgeable tissue culturists spend a part of each day monitoring the cultures in the growing room. The person who monitors the cultures should coordinate the media making and transfer room schedules to meet production requirements. The coordinator can also be the trouble shooter who identifies problems and tries to solve them.

Careful attention must be paid to meeting the lighting and temperature requirements of the particular plants being tissue cultured. Most cultures will do well under the conditions of light and temperature described in Chapter 4, and fortunately, cultures seem to have a wide tolerance for light intensity. In general, plants in culture have low rates of photosynthesis—probably because of the sucrose in media. The fact that cultures are green indicates the synthesis of chlorophyll, which can be an indicator of adequate light. There are exceptions, of course: cultures that do better in higher or lower light intensities, or even in darkness for one or more stages of culture. Carnations (*Dianthus*) and orchids, for example, are often cultured in continuous light, whereas *Hosta* and *Lilium* cultures can multiply in continuous darkness. *Gerbera* responds best to high light intensity (1000 f.c.) during the rooting stage.

DEALING WITH PROBLEMS

Contaminants

Typically the contamination rate in a well-run commercial growing room can be expected to be about 1%. Obviously, as the contamination rate increases, the cost effectiveness of the operation decreases. Contaminated cultures should be removed as soon as they are discovered. Despite the fact that culture containers are almost completely sealed, contaminants can spread to neighboring vessels. Chapter 11 outlines in more detail the types and sources of contaminants commonly encountered in tissue culture, as well as methods for detecting them.

It is easy to tell when cultures are overrun by bacteria, yeasts, or molds growing on the surface of the agar. Other contaminants that grow as a cloud or haze within the medium, often surrounding the submerged base of the culture, are more difficult to see, especially if the agar is not perfectly clear. The best way to detect contaminants within a medium is to look through the medium while holding the test tube up to a light. Culture mites are another specter; they travel invisibly on air currents, carrying with them bacteria and spores. They sometimes can be detected by tiny "footprints" across the medium. Bacteria are most easily observed when Gelrite or other clear agar is used because many bacteria grow below the agar or gel surface, especially if they are emitted from within the stem through the lower cut end.

The best course of action against any contaminant is to sterilize and discard all contaminated cultures immediately upon discovery. Re-examine sterile procedures and check for possible sources of contamination. All reasonable steps must be taken to prevent the spread of existing contaminants and the introduction of new ones. One method of protection that can be effective is to place each tray of containers within a large, sealed plastic bag.

In general, antibiotics are ineffective against culture contaminants, and they can be toxic. Occasionally, however, successes with antibiotics are reported, and they are required in some

Figure 9-1. Growing room shelf and lighting.

cell cultures. One study used a combination of 25 mg cefotaxime, 6 mg tetracycline, 6 mg rifampicin, and 25 mg polymyxin B, per liter. Because these antibiotics degrade if they are autoclaved, they must be cold-filter sterilized before adding to sterilized media. They are also used as a soak in explant cleaning. Although by no means a cure-all, antibiotics are effective against at least some woody plant bacteria (Young et al. 1984), especially when used both as a presoak and in liquid media. Gentamicin sulfate, streptomycin sulfate, and penicillin-streptomycin solution have also been credited with some success. Gentamicin and streptomycin can be autoclaved and still be useful in some instances.

A special method of testing for the presence of contaminants, called indexing, is employed by some tissue culturists. Indexing is particularly valuable to reveal systemic (internal) microbial contaminants in newly started cultures. Indexing for biological contaminants is discussed in Chapter 11.

Media problems

A careful monitor must watch for bleeding, the phenolic exudates from the cultures that color the medium an inky purplish black. This condition can be a result of dull knives, old cultures, or too liquid a medium. Bleeding usually has an adverse effect on the culture, turning the base black and retarding its growth and multiplication. It is worthwhile to transfer such cultures as soon as the symptom is observed.

Other culture conditions to look for are changes in color or growth patterns. It is usually an indication that it is time to vary the formula or environment when leaves turn yellow or brown, when cultures grow too fast with a watery succulence (vitrification), when stems turn red, or when unwanted roots or callus are produced. If cultures produce roots prematurely (in Stage II), the plantlets tend to mature and will cease to multiply. This calls for quick action, or else production will be curtailed. Transferring to media with higher cytokinin may be sufficient to reverse this trend and bring back multiplication. Adjusting formulas to accommodate the special needs of cultures is an art that is necessary for success.

Because of the dramatic influence hormones have on culture growth, the prime suspect to investigate when problems arise is the cytokinin/auxin ratio in a formula. A methodical way to establish the optimum cytokinin/auxin ratio is to draw a grid and plot an auxin and a cytokinin on the 2 axes in increasing increments (Table 9-1). For shoot cultures in the multiplication stage (Stage II), rule out ratios where the auxin content is greater than the cytokinin content (at least in the initial trials). A preliminary test might include the 6 ratios underlined in Table 9-1. To conduct the test, mix media of each extreme ratio, a few other ratios, and the standard formula for comparison. Grow at least 10 tubes in each variation, preferably through 2 transfer cycles. Sometimes the cultures grown in the extremes will show dramatic differences, possibly a helpful clue. For example, a 0/0 ratio (no hormones) has been known to promote elongated shoots where short shoots had been a problem. Zero hormones will not be the entire answer in that case, but it indicates that the problem is hormonal and that less cytokinin (or a different one) should be tested.

Table 9-1. Grid of Cytokinin/Auxin Ratios

Cytokinin (mg/L)

Auxin (mg/L)	0	0.5	1	3	5	10
0	<u>0/0</u>	0.5/0	1/0	3/0	5/0	<u>10/0</u>
0.5	0/0.5	0.5/0.5	1/0.5	<u>3/0.5</u>	5/0.5	10/0.5
1	0/1	0.5/1	<u>1/1</u>	3/1	5/1	10/1
3	0/3	0.5/3	1/3	3/3	<u>5/3</u>	10/3
5	0/5	0.5/5	1/5	3/5	5/5	10/5
10	0/10	0.5/10	1/10	3/10	5/10	<u>10/10</u>

When trying to improve culture response, there are other ways to alter formulas besides varying the cytokinin/auxin ratio. In fact, the number of possible modifications to media is infinite. One can try different growth regulators at different concentrations. Stock solutions can be added to media at ¼-, ½-, or 2-times strength and the results compared with those of the standard strength. Or try varying the individual salts or vitamins. Try various brands, sources, quality, and quantity of agar. Agar concentration, pH, light, and air circulation are other variables to consider. Whenever a formula modification is attempted, always run a few controls (standard conditions that are causing the problem) for comparison. Trials such as these can indeed be trials—time consuming, expensive, and rarely rewarding, but oftentimes necessary.

Cultures in rooting media often benefit from the addition of charcoal, which appears to help remove residual cytokinin. The darkness of charcoal in the rooting area may also be a beneficial factor. The presence of charcoal in rooting media makes it easy to distinguish rooting media from multiplication media, but observation of root formation is more difficult. Charcoal has also been used remedially for some problems in the multiplication stage.

Air requirements

The need for air exchange in *in vitro* culture vessels is generally not a common problem. Most culture vessel systems appear to provide sufficient exchange for normal growth. Some cultures, however, such as potatoes *(Solanum tuberosum)*, are quite sensitive to a shortage of air. In relatively air-tight containers, potatoes exhibit spindly growth and tiny leaves, symptoms not exhibited by rhododendrons, begonias, berries, and numerous other plants grown in the same type of containers and in the same environment. An easy solution is to bore about a 0.25-in (0.5-cm) hole in the polycarbonate disc covering the jar and install a piece of nonabsorbent cotton in the hole before sterilizing. This small allowance for air exchange can make a dramatic difference. Some commercially available covers that allow air exchange are Magenta B-Caps, which are vented caps that fit baby-food jars, vented closures for Magenta GA-7 vessels, and Phytacon vented lids, which fit Phytacon polypropylene culture vessels (Sigma Chemical).

Other methods to help aerate cultures involve pumping sterile air to cultures. Open culture containers may be placed in plastic bags into which sterile air is pumped. Alternatively, an intricate tubing system can supply fresh sterile air to each jar. The advantages of aeration are less vitrification and greater ease of acclimatization.

Thomas and Murashige (1979) examined culture vessel air for 5 gases using gas chromatography. Culture vessels with proliferating shoots had less carbon dioxide (CO_2) than those with callus or nodal explants. The implication was that proliferating shoots had greater photosynthetic activity. Ethane quantities were usually the same as in room air. Acetylene and ethyl alcohol were virtually absent from shoot cultures. Ethylene production was quite variable, but usually higher in callus cultures.

Malcolm Wilkins (1984) compiled a long list of plant responses to ethylene. Some that might apply to shoot cultures *in vitro* are seed and bud dormancy, inhibition of shoot elongation, promotion of radial growth, respiratory changes, epinasty (leaves bend down, short nodes), root and root-hair initiation, tissue proliferation, and curtailment of apical dominance. Under certain conditions, most of these responses can also be induced by auxins; auxins, CO_2, and stress all stimulate ethylene production. Another study reported restricted growth of carnations *(Dianthus)* in sealed vessels and in controls when ethylene was added (Mele et al. 1982). Ethylene is suspected of causing the small leaf size in potato *(Solanum tuberosum)* cultures in tightly sealed jars. Tomatoes *(Lycopersicon)* are reportedly indicator plants for ethylene (Ball 1985). In the presence of ethylene they exhibit epinasty.

Vitrification

Vitrification is a physiological condition wherein tissues appear water soaked or translucent. Vitrified tissues lack chlorophyll and become brittle, taking on a crystalline appearance. Other characteristics are rapid and abnormal growth, short internodes, thick curly leaves, loss of cuticle, excessive ethylene production, and death. Vitrification is a common problem in micropropagation.

The basic cause of the condition lies in the *water potential* of the culture, that is, the water available to tissues. Vitrification is promoted by the type and concentration of agar and cytokinin in the medium. Weak agar and high cytokinin encourage vitrification. Chemically, the problem has been attributed to a deficiency of lignin, which is essential to the structure of cell walls and other tissues (Phan and Hagadus 1986).

Several remedies have been suggested, but the best approach is prevention. Use a minimum of cytokinin and a maximum amount of agar, depending on the extent of the problem and the requirements for optimum growth (Pasqualetto et al. 1986). Maximizing the air exchange through the container closure also helps prevent vitrification (Miles 1990). The addition of phloroglucinol to media has been useful as a remedy or deterrent (Phan and Hagadus 1986). Cold treatment has been found to have some remedial effect as well (Boxus 1978).

Refrigeration

There are, of course, many other problems to contend with in addition to determining the correct medium for a given cultivar. For example, what happens when the transfer crew has not kept up with production demands and a thousand containers are behind schedule? A walk-in refrigerator will do much to alleviate the problem. Chilling can ease an emergency situation in production. If plantlets are ready to be transferred to soil by mid-winter, their growth can be slowed down in refrigeration to delay planting until early spring, thus reducing winter greenhouse fuel costs. Some stock cultures can be held almost indefinitely in a refrigerator, with subculturing done only once or twice a year. Certain cultures cannot tolerate cold storage, but those that can may be placed in it at any stage. Use only samples at first and check them frequently until you learn if the plant can tolerate cold storage and for how long. The temperature in refrigerators should be held at about 35°F (2°C) without light.

Cold storage is also useful for administering chilling treatments. Pear *(Pyrus)* and apple *(Malus)* cultivars respond with additional growth when chilled for 1000 hours prior to greenhouse planting.

SYMPTOMS, CAUSES, AND SOLUTIONS

From years of experience growers know how to deal with many of the requirements and idiosyncrasies of particular plants growing in a greenhouse. Even so, plants do not always perform as expected or in the same way from one grower to another. Significantly less experience has been gained with tissue culture practices, so a grower must be prepared to vary media or other factors as problems arise.

As with most disciplines, there are many problems for which no specific solutions are known. The following list of problems and possible solutions (Table 9-2) is not intended to say, "if you do A and B to solve C, then D will happen." This list will suggest other ideas about the causes of problems and their potential solutions. You and others will help fill in the blanks in this ever-growing, ever-challenging field. For example, the influence of container size on material is a subject of controversy among growers. It is easy enough to test the response of a culture to various containers.

In a larger laboratory most of these problems would be referred to a research and development department. It is the duty of the person or group of research technologists in this department to trouble shoot. They are usually assigned routine tasks, such as maintaining stock cultures, quality control, sanitation, and safety.

Whether or not you have a research department, it is important to record the response of cultures to different media. The results of every trial should provide information on which future operations are based. Maintaining good records justifies the time invested and provides a key to future opportunity. Note both positive and negative results to avoid repetition of errors and to guide future protocol.

As problems arise you will want to do the following:

- Examine laboratory records to learn when a problem started. If other cultures have exhibited the same problem, check the medium history to learn what previous responses have been. Check the original formula to see if the current ingredients agree, or if any changes were made, whether they were beneficial.

- Check all instruments to be sure they are functioning properly, particularly the pH meter, sterilizer, and water purifier.
- Try variations of medium, light, and temperature.
- Study the greenhouse and field habits and requirements of the plant in question to see if these factors yield any clues for treatment in culture.
- Consult outside sources of help, including:
    Computer networks
    Professional organizations
    Tissue culture supply companies
    Cooperative Extension Service, U.S. Department of Agriculture
    Other tissue culture laboratories
    Universities, botany and related departments
    Libraries

Table 9-2. Common Tissue Culture Problems and Possible Solutions

Symptom	Possible Cause	Possible Solution
Explant dies	Disinfectants too harsh	Use weaker disinfectant
	Medium too strong	Use weaker medium ($\frac{1}{2}$ or $\frac{1}{4}$ strengt
	Timing	Obtain explants at different times
Culture blackens and dies	Contamination	Discard culture; review sterile technique and sanitation
	Bleeding	Transfer immediately and more often
	Water problem	Check water purity
	Wrong formula	Change medium
	Too hot	Reduce temperature
Explant live, but no growth	Dormant	Chill for a month
		Start explant at different time
	Needs more time	Transfer explant
		Patience
	Medium too strong	Use weaker medium
	Wrong formula	Change medium
		Decrease salts and hormones
		Layer
		Add charcoal
		Add GA_3 (gibberellic acid)
	Too hot or cold	Change temperature
Growth too slow	Wrong medium	Change medium
		Patience
	Too hot or cold	Change temperature
Shoots too long and spindly	Too little cytokinin	Increase air and/or cytokinin
		Run cytokinin/auxin grid
Poor shoot multiplication	Too little cytokinin	Increase air and/or cytokinin
		Run cytokinin/auxin grid
Shoots too short	Too much cytokinin	Decrease cytokinin
		Run cytokinin/auxin grid
PNo multiplication	Too little cytokinin	Increase cytokinin
		Run cytokinin/auxin grid
	Needs chilling	Cold store 4 to 8 weeks
	Too cold	Increase temperature
	Requires dormancy	Cold store 3 to 8 weeks

mptom	Possible Cause	Possible Solution
t stems, small leaves	Too much cytokinin	Decrease cytokinin
		Run cytokinin/auxin grid
ıwanted callus	Wrong hormones	Decrease cytokinin
		Run cytokinin/auxin grid
		Omit auxin
		Try 1 to 5 mg TIBA per liter of medium
lorotic leaves	Contamination	Check or index for contaminants
	Too hot	Reduce temperature
	Wrong formula	Change medium
		Vary nitrogen
trification	Osmosis upset	Reduce temperature
		Increase air
		Increase agar
	Too much cytokinin	Decrease cytokinin
	Lack of lignin in cell walls	Add phloroglucinol
	Wrong agar	Change agar
	Culture too old	Transfer more often
emature rooting	Wrong hormone balance	Increase cytokinin
		Run cytokinin/auxin grid
ed stems	Stress	Change light and temperature
	Too much sugar	Decrease sugar
	Too little nitrate	Increase nitrate
	Culture too old	Transfer more often
age IV plants die	No roots	Grow roots in culture
		Apply auxin to base
	Too hot, too dry	Adjust humidity and soil moisture
	Pathogens	Test for pathogens
	Wrong soil mix, pH, moisture, or food	Research and experiment
	Too rapid hardening	In Stage III medium: decrease sugar, open jars 1 to 2 weeks
›or multiplication	Senescence	Shorten time in culture
		Increase or change hormone
		Add charcoal
	Pathogens	Test for pathogens
berrations	Medium	Change medium
	Lack of air	Provide sterile air (put filter in cap)

Chapter 10: Hardening Off

Most growers are well acquainted with the hardening-off, or acclimatization, requirements in the greenhouse for seedlings and cuttings. It is much easier to grow seedlings than it is to establish tissue cultured plantlets in greenhouse conditions. If care is not exercised during this very critical stage, high losses can result. An abrupt change to the lower humidity and greater light of the greenhouse can be fatal to plantlets in a very short time. But if you employ good techniques for acclimatizing tissue cultured plantlets, the effort spent developing the cultures to this stage will have been worthwhile.

Ultimately the quality of Stage IV *(in vivo)* plantlets depends on the quality of the plantlets that come from the growing room. Oftentimes Stage III (the rooting stage *in vitro*) is omitted and plantlets are rooted in soil directly from Stage II. The success of this practice depends on the particular plant, its ease of rooting, and on the grower. Sometimes the roots that develop in Stage III are not functional in soil, in which case the plantlets may as well go to soil from Stage II. The advantages of skipping Stage III are less labor costs, reduced transfer room use, and more space in the culture growing room. Root systems are also generally stronger if Stage III is bypassed. In one report (Mikkelson and Sink 1978), the production cost of *Begonia rex* was cut by 50% when direct rooting was employed. But the plantlets of some species may require Stage III for more rapid rooting, better height, more sturdiness, or better conditioning.

Acclimatization can be started while plantlets are still *in vitro*, or it can wait until Stage IV when the plantlets are moved from containers to soil. If microcuttings are to be taken in Stage II, then the hardening off will occur entirely in Stage IV.

A major problem faced in acclimatization is water loss. Plants have *stomata*, small openings located between 2 crescent-shaped sunken guard cells (see Figure 2-3), that serve to control water balance and allow gas exchange for photosynthesis and respiration. They are found primarily on leaves and, to a lesser extent, on stems and other structures. Many plants have stomata only on the undersides of their leaves, but some have stomata on both surfaces. In the process of *transpiration*, water vapor escapes through the stomata and *cuticle* (a waxy layer covering the epidermis that protects plants from excessive water loss) to the surface, where it evaporates. Normally, stomata open and close in response to turgidity. When a tissue cultured plantlet is transplanted to soil, the stomata usually are fixed open until they are able to adjust to the lower humidity—tissue culture containers generally have high humidity—and greater light, or until new leaves are produced. This period varies with the particular plant, its conditions in culture, and its new environment.

Electron microscopy has revealed atypical characteristics on leaves that develop in tissue culture. Often there is an abnormally high number of stomata per given surface area. The unusual nature of epicuticular wax (a membranous covering) on leaves and stems in culture has also been observed. It was found to be abnormal in both its chemical make-up and its appearance (Preece and Sutter 1991), and the transpiration protection normally afforded by this waxy coat is depleted in cultures. Further water loss in tissue cultured plants results from *guttation*, the exudation of water from *hydathodes* (organs connected to the veins of leaves and situated at the tips of the leaves or at the vein ends on the leaf margins).

In addition to these changes in morphology and transpiration, the entire photosynthetic process is confused in *in vitro* plantlets. In culture media, plantlets make little use of carbon dioxide (CO_2) because they draw on the sucrose in the medium for energy instead of depending on photosynthesis. In order to encourage photosynthesis in plantlets, CO_2 can be introduced into culture vessels or the greenhouse. This is done most readily in an acclimatization chamber rather than while the plantlets are still in culture. If CO_2 is to be introduced, there must be a corresponding increase of light for the CO_2 to be effective. One study found that several woody plants, including *Kalmia*, lilacs *(Syringa)*, grapes *(Vitis)*, apples *(Malus)*, and blackberries *(Rubus)*, benefited significantly when CO_2 was introduced into an acclimatizing chamber (Mudge et al. 1992). In this study, optimum irradiance from fluorescent fixtures was at about 130 µmol m^{-2} sec^{-1} (about 1000 f.c. or 10,000 lux) for *Kalmia* and slightly higher for *Syringa*. Carbon dioxide levels were at 1200 parts per million, monitored with a CO_2 monitor/controller. A household-type humidifier provided humidity.

Other characteristics exhibited by tissue cultured plantlets that are deviations from normal plants can include thinner roots and stems, thinner leaves, underdeveloped palisade layers, fewer trichomes (hairs, scales, etc.), less collenchyma, reduced vascular tissue, and poorly developed and less chlorophyll.

If plantlets are to be rooted *in vitro* in Stage III, there are several ways of preparing for acclimatization. Small bags of desiccant (silica gel) can be hung in the containers to lower the humidity. Watch this option carefully because the cultures may dry too quickly. Container covers can be placed to fit more loosely and allow more water vapor to escape, but if left for more than a few days contaminants may build up, depending on the purity of the room. The survival rate of carnations *(Dianthus)* approached 90% when containers were left uncapped for 9 days in a room with 50 to 70% humidity before transplanting to soil (Ziv 1986). Others have used this practice successfully both indoors and in greenhouses. Depending on room humidity, it is best to remove the lids gradually. Another effective method is to use lids with filters, which allows some desirable air exchange and can be applied much earlier than removing the lids. Bottom cooling to reduce relative humidity in container head space has worked effectively for *Cynara*, *Iris*, *Rosa*, and *Maranta* (Maene and Debergh 1983).

Using a stiffer agar will reduce available moisture somewhat, but it will make it more difficult to separate the roots from the agar and can cause roots to break off when they are handled. Harder agar also reduces the availability of certain nutrients. If a liquid medium is used for Stage III, many more root hairs will develop, but the roots may be less sturdy than if grown in agar and they may not adapt well to soil.

Decreasing salt levels, especially nitrates, often helps to induce rooting. The amount and type of cytokinin used in Stage II can also affect rooting. *Liriope*, *Schefflera*, and *Philodendron* have been found to survive transplanting better by 80% when 2iP or kinetin are used instead of BA in Stage II. In Stage III, cytokinin is usually eliminated and the auxin level raised. The auxin level is very important, and too much auxin can be worse than none at all. Sometimes 2 auxins used together work better than either one used independently. Lowering sucrose levels in Stage III may encourage photosynthesis, but raising it may favorably affect the water potential. Phloroglucinol has proven beneficial in rooting of some fruits.

Although lighting requirements will vary widely, many cultures root better in increased light (350 to 600 f.c.). On the other hand, some bulbous and rosaceous plantlets form root initials more readily in darkness than in light.

Common procedure for transferring plantlets from the culture growing room to the greenhouse is to wash off the agar from the roots, plant the plantlets in artificial soil in undivided planter trays (11×22 in [28×55 cm]), and place them in high humidity in tunnels or tents on benches in shaded greenhouses. Over a 2- to 4-week period, the sides of the tent should be gradually opened (Figure 10-1) and the amount of mist gradually reduced to lower the humidity, thus allowing the existing leaves to adjust and/or assisting new leaves to grow. These enclosures are variously equipped with mist, bottom heat, exhaust fans, cooling pads, and lighting with appropriate timing and light-sensing devices. An entire greenhouse might be similarly equipped in lieu of a tent, but the larger the greenhouse, the more difficult it is to control moisture and humidity.

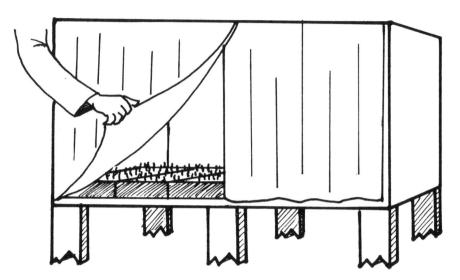

Figure 10-1. Plastic-covered greenhouse bench.

A fog system is an ideal way of maintaining high humidity without over saturating the plants you are trying to harden off—although fog is expensive and difficult to maintain when exhaust fans or vents are working to reduce heat. Humidifiers are less expensive and will do the job well, especially in smaller areas. Misters are also commonly used—the finer the mist the better. Misters need to be controlled by a timer and watched carefully so that the plants will not become saturated. Greenhouse shade can be maintained at 50%. If desired, the photoperiod (day length) may be extended, with artificial light as low as 200 f.c.

A growing room equipped with fluorescent lights over shelves, similar to the culture growing room, is a good option for establishing tissue cultured plantlets (from either Stage II or Stage III) into soil. Heat and light are easily controlled in this scenario. Humidity can be maintained by planting in 11 × 22 in (28 × 55 cm) seedling trays with plastic covers, called Humidomes (Figure 10-2). Smaller containers can also be used; a transparent 4 × 4 in (10 × 10 cm) plastic berry cup can hold up to 25 plantlets. A second cup inverted over the top and taped makes a miniature greenhouse (Figure 10-3). Drainage holes in the inverted top cup should allow room for a wash-bottle stem to be inserted for watering (but to start with the holes may need to be taped over). They can also be watered from the base by flooding in a water-tight bench. The smaller containers will help confine any disease problems to fewer plants. It is best to not crowd the plantlets; the seedling trays ideally should house about 200 and the cups about 20 microcuttings or plantlets.

One of best protective covers is a non-woven cloth of spun bonded polyester, one brand of which is Reemay. An open-bottom tray inverted over a tray of plantlets will serve as a holder for this cover while allowing light and water through. It prevents droplets from falling on the plants, holds humidity and moisture, and modifies temperature.

As plantlets are moved from Stage II or Stage III to soil mixes, it is desirable to wash off as much of the agar as possible because molds, yeasts, bacteria, and insects thrive on the nutritious

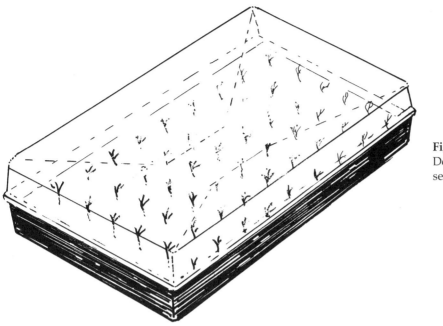

Figure 10-2. Dome-covered seedling tray.

agar. Various fungicides should be tested on a few plantlets to learn which ones can be used safely; always carefully follow the instructions on fungicide labels. Avoid using fungicides unless it is absolutely necessary. Some growers use a rooting-hormone powder or auxin dip, such as Rootone or Dip-n-Grow, at this point. Another approach is to transplant shoots to pots of soil mix that are then placed in jars and covered as they had been in culture.

One walnut grower in California has developed a patented process that involves pretreatment in culture, root induction in flats in a growing room, and field establishment without a greenhouse stage (Driver 1985). The root-ready plantlets in the field are first covered with plastic drinking cups, which in turn are covered with 2 foam cups. The cups are removed in 3 stages. This method was used successfully with *Juglans paradox* and *Prunus persica*.

The use of antitranspirants to help acclimatization is not widespread because research has

Figure 10-3. Clear plastic berry cup with cover.

been unsuccessful or inconclusive. Minor success was achieved with the use of Folicote on apple *(Malus)* (Hutchinson 1984).

There are probably about as many artificial soil mixes as there are growers. Soil media are not always defined in the literature, which slights this very important aspect of hardening off. Some of those used for micropropagated plants include the following:

Fine charcoal/fine redwood (or fir) bark/ perlite 1/1/1	Orchids
Jiffy 7's (peat pellets that expand in water)	*Dianthus, Rubus*
"Mica-peat" + NAA solution	*Tsuga*
Milled sphagnum moss	Top dressing for *Rhododendron*
Peat moss	*Kalmia*
Peat moss/bark/sand/sawdust 1/1/1/1	woody plants
Peat moss/perlite 1/1	*Betula, Malus, Rhododendron, Rubus*
Peat moss/perlite 1/2	*Rhododendron*
Peat moss/perlite/sawdust 3/2/1	*Rhododendron*
Peat moss/perlite/sawdust/pumice/ vermiculite 3/2/1/1/1	woody plants
Peat moss/perlite/vermiculite 1/2/1	*Acacia, Larix, Pinus*
Peat moss/pumice 1/1	*Pinus radiata*
Peat moss/sand 1/1	*Vitis*
Peat moss/sand 1/1.5	*Pinus taeda*
Peat moss/sand/perlite 3/1/4	Fruit-tree rootstocks
Peat moss/vermiculite 1/1	*Alnus, Salix, Typha*
Vermiculite/sand 1.5/1	*Pinus taeda*

Chapter 11: Dealing with Biological Contaminants

This chapter describes some of the primary methods used to detect and identify microbial and viral contaminants. Topics related to contamination selected for discussion are

- Laboratory sanitation as it relates to solid surfaces, air, and humans.
- Biological quality-control procedures for monitoring sanitation.
- Detection and partial identification of external and systemic bacterial and fungal contaminants using plant tissue indexing procedures.
- Detection and identification of viral contaminants using ELISA (enzyme-linked immunosorbent assay), an immunological procedure.

The term "biological contaminants" refers primarily to bacteria and fungi found on or within explants or in the laboratory. Sources of contamination in the laboratory include laboratory air or solid surfaces, humans, and improperly prepared tissue culture media. Biological contamination of explant tissue is thought to originate primarily from either a systemic contaminant or from improper use of aseptic laboratory procedures (Cassells 1991). Occasionally, external contaminants present on explant tissue, such as bacterial endospores, may be resistant to chemical sterilization. An endospore is a highly resistant, dormant asexual cell that forms in some microbes. When placed in a favorable environment, it germinates into an actively growing vegetative cell.

Plants normally excrete substances from their roots that stimulate microbial growth in the *rhizosphere* (the soil surrounding plant roots). Microbes can enter plant tissues through natural openings or wounds, from where they begin to colonize the plant's tissues, or they can enter plant tissues with the assistance of *vectors* (disease carriers) such as insects (Tarr 1972; Matthews 1981). Plants carry a wide range of microflora, including procaryotes (bacteria and bacteria-like organisms), fungi, algae, viruses, and viroids (infectious agents smaller than viruses and composed only of RNA) (Campbell 1985). Viral contaminants do not pose a problem when located outside of a plant; only after they infect the plant can viruses use the plant's biochemical systems for viral reproduction.

Contamination in tissue culture operations can be costly, particularly if the contamination rate is greater than 1%. If contaminants are observed in test tubes or jars, the containers should not be opened and should be sterilized before discarding the contents. Many plant tissue culture managers, however, do nothing more than make a casual observation of contaminants. If the agar is clear and no foreign growth (such as bacteria, molds, or yeasts) is observed and if the cultures look healthy, they are presumed healthy. All too often, no further action is taken due to the time and expense involved, lack of interest, lack of information, or a willingness to gamble that unseen contaminants either do not exist or will not cause any problems. But if the propagator is seriously concerned and is dedicated to having the cleanest possible plants for sale or for in-house use, then further tests are necessary to determine the origin and nature of the microbial contaminants and to better avoid any such future problems.

Determining the source of microbial contamination in tissue culture containers is no small task. Contaminants are usually first observed after some days or weeks of incubation in the growing room when containers are taken back and forth from the transfer room to the growing room. Contaminants have a much faster rate of growth than plant tissues, so when they populate a container they will eventually overrun and destroy the culture.

LABORATORY SANITATION

By way of definition, *sanitation* is the process of substantially reducing, and then maintaining, the microbial population in the air and on objects in the laboratory to acceptable levels; in short, being clean by using water, soap, or detergent. *Disinfection* implies the use of chemical agents to kill pathogenic microbes, without necessarily sterilizing the material to which the chemical is applied. *Sterilization* is the destruction of all living matter. Sinks are cleaned or disinfected, explants are disinfected, but media and transfer tools are sterilized.

A good sanitation program for the media preparation, transfer, and growing rooms minimizes the need for quality-control monitoring. If an enviable 1% or less of containers in the culture growing room have contaminants, it is an indication that the laboratory has good sanitation procedures. If contamination is significantly higher than 1%, then it is desirable to initiate procedures for isolation and identification of external and systemic microbial contaminants.

Possible sources of contamination are air from the transfer hood HEPA filter, various solid surfaces, including tools and containers present in the transfer hood, and flaws in the sterile technique used by the person transferring cultures. An exercise in the proper use of sterile technique is detailed in Chapter 8. Anyone not already competent in sterile technique will benefit by first practicing these procedures before doing critical work. To do so will save time and money.

Practical knowledge of sanitary techniques is necessary for successful execution of various laboratory procedures. A daily washing and rinsing of laboratory floors and bench, cart, and table surfaces is essential. Sweeping the floors, other than for spillage reasons, should be avoided because it stirs up dust particles. A detergent solution containing a surface tension depressing agent (such as 1% Tween 20) is adequate for washing of laboratory surfaces. After washing, surfaces should be rinsed and mopped with clean water to speed drying. The walls should be washed at least once every 2 weeks, and the ceiling at least once per month.

Human element

Studies have shown that the number of bacterial colonies found per liter of air rises in direct proportion to the number of people in a room (Nester et al. 1995). This would suggest that there are advantages to having a laboratory with few personnel.

Some technicians tend to shed a large number of microorganisms from their skin. Such shedders, especially, should wear gloves, head covers, and perhaps masks. If properly positioned, faces and hair should not be over containers or even in the hood. A shield or other barrier at forehead height can serve as a reminder to keep well back from open containers. Coughing or sneezing results in the release of many infectious particles. Swab samples taken from hands and arms can be grown as described earlier in the chapter for solid surfaces. One test that can illuminate the importance of cleanliness is to place your fingers on agar media in petri dishes before and after hand washing. After incubation, compare the colonies growing in each of the dishes. Careful prewashing of hands and exposed arms with a disinfectant soap or alcohol is recommended for all those involved in tissue culture transfer.

It should be noted that most contaminants encountered in tissue culture operations are of little or no risk to healthy people, provided standard microbiological procedures of containment are employed. The few organisms posing a slight risk are usually normal body-inhabitant species. If you have a special health problem you should be especially diligent in following the precautions listed. Examples of such conditions are diabetes, allergies (especially to fungi), and disorders of the immune system. In order to maintain safe conditions in a tissue culture transfer room, common sense and strict personal hygiene should prevail.

TESTING FOR SOLID-SURFACE BIOLOGICAL CONTAMINANTS— STREAK PLATES

A quality-control test for detecting and enumerating microbial contaminants on solid surfaces is the cotton swab–streak plate procedure. In this test, swab samples are taken from various solid surfaces in the laboratory and tested on microbial-enrichment agar media that encourage any microbes present to form colonies, thereby enabling their enumeration and subsequent identification. (A colony is a visible population of cells that arises from a single cell.) Identifying the kinds of colonies is often helpful for tracing the source of contamination. The number of colonies found provides an approximation of the number of microorganisms present per unit of surface area. If the test indicates a large number of contaminants on one solid surface area and only a few on another area, additional sanitation of the highly contaminated area may be necessary. It may be advisable to repeat the test.

The common microbial-enrichment agar media for detection of contaminants are trypticase soy agar (TSA) and Sabouraud's dextrose agar (SDA). TSA medium is used to detect bacteria and SDA medium is used to detect fungal contaminants.

Equipment and supplies
> Sterile microbial-enrichment agar media
> 1-liter Erlenmeyer flasks
> Distilled water
> Aluminum foil
> Autoclave or pressure cooker
> Sterile petri dishes (plates), 100×10 mm
> Labeler
> Sterile wooden-stick nonabsorbent cotton swabs or dacron tip swabs, tips
>   approximately 1 cm long, in capped test tubes
> Test tube rack and basket
> Test tubes containing sterile water, capped
> Glass-marking pencil
> Inoculating loop and handle, approximately 20 cm (8 in) long
> Bunsen burner or Bacti-Cinerator
> 25°C (77°F) and 37°C (98.6°F) incubators (if available)
> Wastebasket

Place the individual cotton swabs in test tubes, cap, and sterilize in autoclave or pressure cooker. Various cap types, including nonabsorbent cotton plugs, screw caps, and push-on caps, are available.

PREPARING MICROBIAL-ENRICHMENT AGAR
> Prepare about 20 ml of the microbial-enrichment agar solutions in the petri dishes as follows:
>   1. Prepare TSA medium (commercial dehydrated mix) in a flask using 45 g of mix
>   per liter of distilled water. For SDA medium, prepare using 65 g per liter.

2. Cover the flasks with aluminum foil and sterilize in an autoclave for 15 minutes at 250°F (121°C).

3. After processing, cool the freshly autoclaved agar to approximately 115°F (46°C)—cool enough that the flask of agar can be held to your cheek without burning your cheek. Agar will solidify at approximately 105°F (40°C).

4. To pour the agar from the flask, partially lift the petri dish cover using sterile technique (see Figure 11-1b) and pour the agar into the plate. If the agar is too hot when poured, moisture will condense on the underside of the lid.

If the petri dishes are to be stored or the agar cannot be poured before it hardens, the agar will require remelting. If storage is necessary, when you are ready to remelt the agar, do so in a steamer, or on a hot plate (use caution because it may burn), or in a microwave oven (use caution because it may quickly boil over).

PREPARING SWAB SAMPLES FROM A SOLID SURFACE

1. Delimit a solid surface area (bench top, ceiling, or wall) of 3 square inches.

2. Label each test tube containing a sterile swab to identify it with its corresponding surface.

3. Carefully remove a sterile swab from a test tube, holding the test tube cap as illustrated in Figure 8-1. Replace the cap and return the empty test tube to the test tube rack.

4. Moisten the swab by dipping it in a test tube of sterile water and squeeze out the excess moisture by pushing against the inside wall of the test tube.

5. Swab the area to be sampled using a back-and-forth motion and aseptically return the swab to the same test tube from which it came. When inserting the swab into the test tube, break off the portion of the stick that is in contact with your fingers.

6. Replace the cap and return the test tube containing the swab to the laminar flow hood for streaking on the agar plate.

7. Using fresh swabs for each surface, repeat the swabbing procedure for other surface areas.

PREPARING AN AGAR STREAK PLATE

The purpose of streaking is to dilute the contaminants sufficiently, thereby enabling the formation of individual colonies, each one presumably arising from a single contaminated cell.

1. With a marking pencil, divide the underside of the petri dish containing the agar (as prepared above) into 4 quadrants (Figure 11-1a). On the underside of the petri dish, label the date, the area to be sampled, and the type of medium—TSA or SDA.

2. Aseptically remove a processed swab from a test tube. With the other hand, lift the petri dish cover above the plate and hold it at an angle such that it shields the agar surface from possible contamination during the streaking process (Figure 11-1b).

3. Still holding the petri dish lid, gently touch the swab containing the inoculum to the upper-right surface of quadrant one (Figure 11-1c).

4. Discard the swab into the wastebasket.

5. Using a fresh sterile swab, streak the inoculum. Streak by gently rubbing the inoculum back and forth approximately 20 times over quadrant one (Figure 11-1d).

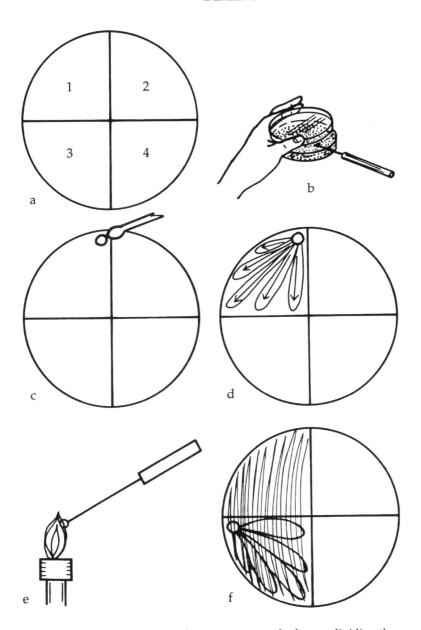

Figure 11-1. Procedure for streaking an agar streak plate: a. dividing the petri dish into 4 quadrants; b. lifting the petri dish cover; c. touching the swab to the quadrant; d. streaking the inoculum; e. sterilizing the loop; f. rotating the petri dish.

6. Replace the petri dish lid and discard the swab.
7. Sterilize the inoculating loop by heating it in a Bunsen burner flame or Bacti-Cinerator (Figure 11-1e).
8. Lift the petri dish lid and cool the inoculating loop by touching it to an uninoculated portion of the agar surface.
9. Rotate the petri dish 90° counterclockwise.
10. Streak the inoculating loop back and forth from the first quadrant to the second quadrant of the agar plate with about 25 streaks (Figure 11-1f). Streak quickly,

letting the loop ride easily on the agar surface. It is all right if the streaks overlap. Too much pressure will break the agar surface, and too little will result in irregular streaks.

11. Streak the contents of quadrant 2 into quadrant 3 using the same procedure.
12. Streak the contents of quadrant 3 into quadrant 4 using the same procedure (steps 7–10).

The above streak-plate procedure should give ample dilution for seeing individual colonies following incubation. If not, you may wish to use an alternative streaking procedure (Figure 11-2), or one of your own modification. Modified procedures may allow for better isolation of single colonies, particularly if the swab sample contains a high concentration of contaminants. One occasional problem is the presence of *Proteus* or other motile bacteria, which rapidly spread over the entire plate, resulting in only a few or no individual colonies.

After the streaking procedure, the petri dish containing the inoculum must be incubated. Incubate the TSA plates at 37°C (98.6°F) for 2 to 3 days and the SDA plates at 25°C (77°F) for 5 to 7 days. The plates should be inverted to prevent any moisture inside the lid from falling on the inoculum. If no incubator is available, store the plates at room temperature for 1 to 2 weeks.

Following incubation, observe the petri dishes for the number and kinds of microbial colonies. If no colonies appear on the plates, one can conclude that the solid surface from which the swab was taken is sterile with respect to microbes able to grow on the selected enrichment medium. Over a period of time, keeping a log of your results will give additional meaning to such studies. For example, you may find that at certain times of the year there is a heavier load of contaminants than at other times. This may also suggest increasing the frequency of the sanitation program during such times.

OBSERVING CONTAMINANTS

For preliminary characterization of microbial colonies that may have developed on your streak plates, observe the following characteristics:

- Color: green, golden, tan, gray, black, white, pink, red
- Size: pinpoint, small, medium, large, spreading
- Shape: round, irregular
- Elevation: flat, convex, crater-like
- Margin: smooth, irregular
- Texture: smooth, rough, woolly, filamentous
- Consistency: mucoid, moist, glistening, dry

A combination of morphological and physiological studies is required for accurate identification of bacteria and yeasts. *Bergey's Manual of Systematic Bacteriology* (Holt 1984) is a useful key for identifying bacteria; *Yeasts: Characteristics and Identification* (Barnett et al. 1991) is a key for identifying yeasts. Filamentous fungi, such as molds, can sometimes be identified by making a study of their colonial (Figure 11-3) and cellular (Figure 11-4) morphology using a conventional 40× dissecting microscope and comparing what you find with a key, such as Barnett and Hunter's *Illustrated Genera of Imperfect Fungi* (1972). Dissecting microscopes with higher magnification (to 200×) are also available. These microscopes have an inverted stage, with the illumination source directed from above the stage. They can also be used for meristems and for observing microscope slide preparations and well plates used in the ELISA test for viruses (dis-

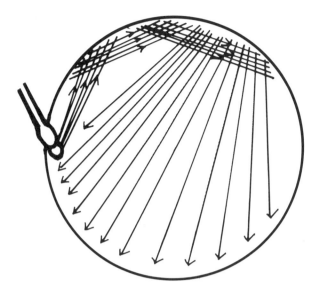

Figure 11-2. Alternate streaking procedure for streak plates.

cussed later in the chapter). A covered slide preparation is sometimes necessary to better observe details of the cellular morphology of filamentous fungi. An introductory laboratory manual, *Microbiology Experiments: A Health Science Perspective* (Kleyn et al. 1995), contains exercises related to all of the above identification procedures.

For normal production quality control, the above procedures for detection and enumeration of contaminants should prove adequate. Further identification of colonies to the genus and species level requires additional training. A good introductory course in microbiology usually includes methods for bacterial identification and sometimes methods for yeast and mold identification.

WET MOUNT SLIDE PREPARATION

It is fairly easy to determine if a non-filamentous colony is a bacterium or a yeast using a wet mount slide preparation for observation under a compound microscope. Refer to Appendix B for information on the proper use of microscopes.

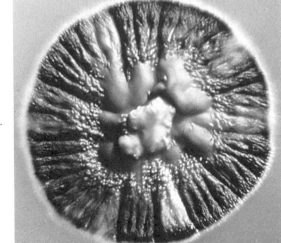

Figure 11-3. A typical mold colony.

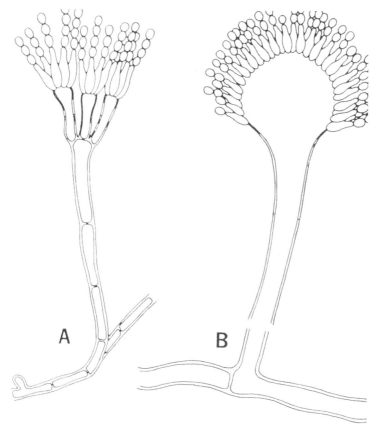

Figure 11-4. Cellular structure of (A) *Penicillium* and (B) *Aspergillus* molds. Reprinted with permission from the Department of Microbiology, University of Washington, Seattle.

Equipment and supplies

>   Microscope slides (75 × 25 mm), 0.13 to 0.16 mm thick
>   Coverslips (22 × 22 mm), 0.13 to 0.16 mm thick
>   Dropper bottle containing distilled water
>   Inoculating needle
>   Compound light microscope
>   Lens immersion oil (high viscosity)
>   Microscope lens paper and lens cleaner

Procedure

1. Place a small drop of distilled water on a clean microscope slide.
2. Place a minute amount of the microbial colony on the tip of a sterile inoculating needle and mix into the drop of water. For best results, a very slight clouding should occur after mixing with the water. The needle must be resterilized after every use.
3. Cover the drop with a clean coverslip. If the coverslip floats on the drop of water, the droplet is too large.
4. Place the slide on the microscope stage and examine it initially with the low-power (10×) objective. When the object is in focus, you should see single cells, and perhaps you will be able to determine whether they are bacteria, which appear very tiny at this magnification, or yeasts, which are much larger than bacteria and

may show some buds attached to the mother cell (Figure 11-5). Sometimes it is necessary to use a stain in order to see bacteria or yeasts.

5. If your microscope is parfocal, switch to the high-power (45×) objective. When properly in focus, you can readily determine if the cells are bacteria or yeasts. If they are bacteria, place a drop of immersion oil on the surface of the coverslip and carefully switch to the oil immersion objective (95×). Now you may see one or more bacterial forms: coccus (spherical; Figure 11-6), rods (cylinders, bacillus; Figure 11-7), or spirals (regularly curved rods).

6. When finished, carefully wipe off the excess lens immersion oil from the oil immersion objective with a piece of clean lens paper.

MICROBIOLOGICAL STAIN

Bacteria are often difficult to observe with a wet mount because their *optical density* (the degree of opacity of a translucent medium) is similar to that of water. Staining will increase the optical density of the bacteria so that their shape and arrangement can be distinguished more easily. Note their shape (rod, coccus, or spiral), their arrangement (coccus may be single, double, tetrad, or cluster), and their size, all of which are important factors for identification.

One objection to staining is that it kills the bacteria, making it impossible to determine motility. Some bacteria have long, whip-like appendages, known as flagella, which enable motility. The flagella are very thin and require special stains to be visible. Some common bacterial stains are methylene blue (also used to stain yeast cells) and crystal violet.

Equipment and supplies

Microscope slides (75 × 25 mm), 0.13 to 0.16 mm thick
Coverslips (22 × 22 mm), 0.13 to 0.16 mm thick

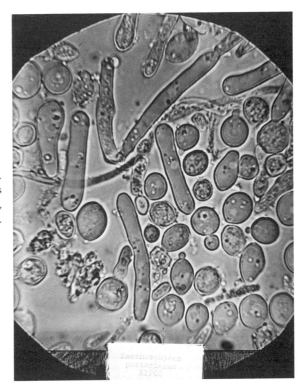

Figure 11-5. Two yeasts and a mold. *Saccharomyces cerevisiae*, with oval cells (some with buds), and *S. pastorianus*, with cigar-shaped cells.

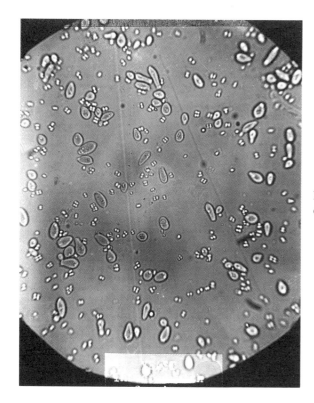

Figure 11-6. A mixture of yeast and a small coccus bacterium, *Sarcina*.

Glass-marking pencil
Dropper bottle containing distilled water
Inoculating needle
Bunsen burner or Bacti-Cinerator
Forceps or clothespin
Prepared crystal violet and methylene blue stains
Blotting paper or paper towel
Compound light microscope
Lens immersion oil
Microscope lens paper

Procedure

1. Prepare a clean slide by washing with a detergent solution. Rinse and dry with paper towel.
2. Using a glass-marking pencil, inscribe a circle on the center of the slide.
3. Place a small drop of distilled water inside the circle.
4. Using a sterile inoculating needle, remove a small amount of a colony and mix it in the water until the drop becomes slightly cloudy. If the broth is too cloudy, uneven staining will occur; if too clear, it may be difficult to find the bacteria. Allow the slide to dry for about 10 to 15 minutes.
5. Attach the bacteria to the microscope slide by heat fixing. This is done by passing the slide through a Bunsen burner flame 2 to 4 times, until the bottom of the slide is lukewarm. A Bacti-Cinerator may also be used. In gently heating the slide, most of the bacteria will adhere to the slide without causing serious distortion of cellular morphology.

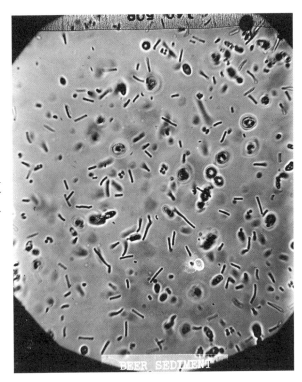

Figure 11-7. Beer sediment, with autolyzed (self-digested) yeast cells, *Sarcina*, and many rod-shaped bacteria.

6. Hold the slide level over the sink using forceps or a clothespin. Add a drop or two of stain inside the circle. Let stand for 20 to 30 seconds.

7. Wash off the excess stain with tap water and dry the stained slide with blotting paper or a paper towel. Take care not to rub the stained area sideways, which could remove the bacteria.

8. Observe the slide using the oil immersion objective of the microscope, following the same focusing procedure outlined above (i.e., start with the low-power objective).

9. When finished, carefully wipe off the excess lens immersion oil from the oil immersion objective with a piece of clean lens paper.

For observing yeasts, the same general procedure is followed.

1. Place a small drop of methylene blue on a clean slide.

2. Using a sterile inoculating needle, remove some growth from a yeast colony and mix it in the methylene blue until the drop is slightly cloudy. Place a clean coverslip over the drop.

3. Observe with the 10× and 45× microscope objectives.

TESTING LABORATORY AIR

Properly filtered inside air should not pose a contamination threat in the laboratory. The threat comes largely from people and their activities within the laboratory. The air in the media preparation room need not be as clean as the air in the transfer and growth rooms. However, in order to maintain the highest cleanliness standards, the air in all 3 rooms should have some positive pressure. The ventilation system of the laboratory should be constantly removing old air and introducing new, fresh air. Periodic testing will ensure that air cleanliness remains high.

PETRI DISH PROCEDURE

1. Remove the lids from a dozen or so petri dishes containing sterile agar media, either MS media or special media such as SDA or TSA for fungi or bacteria, respectively.
2. Place the lids on a sterile surface with the inside down so that no contaminants will drift into them.
3. Place the open petri dishes in various locations throughout the room for various time periods, such as one at each location for 1 minute, 5 minutes, 15 minutes, and 30 minutes.
4. Replace the lids and seal the dishes with tape.
5. Label each dish to identify the medium, exposure time, and location.
6. Invert and incubate in a warm place (77 to 86°F [25 to 30°C]) for a week or two. Petri dishes should be inverted during incubation in order to prevent any moisture present on the lids from falling onto the agar surface, and thus interfering with colony formation.
7. Examine the dishes daily for numbers and kinds of colonies growing on the media.

Usually each colony will originate from a single cell, which multiplies to form a colony large enough to be seen with the unaided eye. If no colonies appear in the petri dishes, you can likely conclude that the air is devoid of living cells (i.e., sterile). This method is especially useful for periodic assessment of the microbial content in the laminar flow hood. The number of colonies found in the hood should be minimal. If the numbers increase significantly, that would suggest a faulty HEPA filter.

ENUMERATION AND DIFFERENTIATION OF TOTAL PARTICLES IN AIR

A membrane filter apparatus attached to a portable battery-operated vacuum pump (Figure 11-8) is sometimes used to assess air quality in the transfer room or the greenhouse. The pump has a vacuum source to collect air contaminants on a membrane filter disk. A direct microscopic examination of the membrane provides a total count of both viable and non-viable particles (Kleyn et al. 1981).

Equipment and supplies

Autoclave or pressure cooker
Membrane filter holder assembly
Sterile membrane filters
Tweezers or forceps
Battery-operated pump, with vacuum source
Tygon hose, 0.25 in (0.5 cm)
Petri dish
Scalpel or knife blade
Microscope slides (75 × 25 mm)
Acetone or lens immersion oil
Compound light microscope

The air-flow rate of a Mine Safety Appliances Flow-Lite pump is adjustable from 0.5 to 3.5 liters per minute. The pump costs about US $580. A pump from SKC Inc. (224-44xR) costs US $435 and is adjustable from 0.005 to 5 liters of air per minute. A Gelman 47-mm Polycarbonate In-Line Filter Holder (1119) runs about US $50 (Figure 11-9), and the filters (0.45 µm

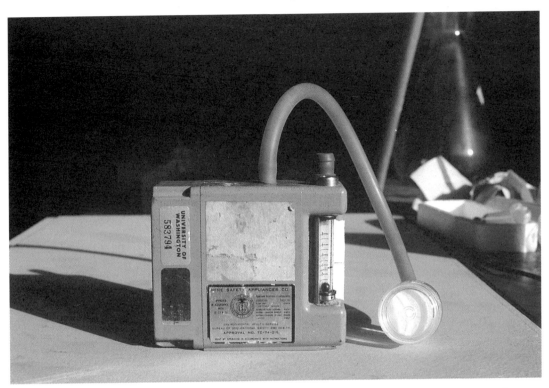

Figure 11-8. A membrane filter holder assembly attached to a portable battery-operated pump for sampling air. Courtesy Mine Safety Appliances Co.

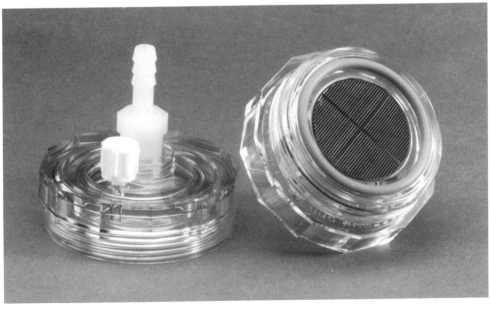

Figure 11-9. Membrane filter holder assembly for air sampling. Courtesy Gelman Sciences Inc.

porosity, 47-mm diameter with grid; Gelman GN-6 Metricel, 66278) cost US $80 for a 200-membrane package.

Procedure

1. Autoclave a wrapped membrane filter holder assembly outlet with attached hose barb adapter, support screen, and inlet (see Figure 11-9).
2. Using sterile tweezers, insert the support screen in the membrane filter holder outlet.
3. Aseptically remove a sterile membrane filter from the package and place it on top of the support screen.
4. Aseptically screw the filter inlet onto the top of the filter outlet.
5. Attach the hose barb adapter to one end of the Tygon hose. Attach the other end of the hose to the inlet fitting on top of the pump (see Figure 11-8).
6. Turn on the pump and adjust to the desired air-flow rate. After a prescribed period of time, turn off the pump. The air-flow rate and duration time of sampling are dependent on the contamination level of the air sample.
7. Using sterile tweezers, aseptically remove the membrane filter from the membrane filter holder and place the filter in a sterile petri dish.
8. With a sterile scalpel, cut off a small portion (about 10×10 mm) of the membrane and transfer it aseptically to the surface of a clean microscope slide.
9. Add a drop of acetone or lens immersion oil to clear the membrane surface, and place the slide on the microscope stage.
10. Using the low- and high-power objectives, examine the slide for particle types (amorphous dust, bacteria, and spores) and numbers relative to one another (Figures 11-10a, b). By examining a number of microscopic fields you can determine the average number of particles per field. This in turn can be related to the number of particles per field per liter of air.

ENUMERATION AND DIFFERENTIATION OF VIABLE PARTICLES IN AIR

The procedure for determining *viable* (capable of germinating, living, growing) particles in the air differs from that of determining the total number of particles in that the viable particles are collected in a container of sterile water, rather than on the surface of a membrane filter. This is done because many viable particles will become dehydrated and die when collected on a membrane surface.

In this process, the container for the sterile water has 2 outlets: one serving as a hose attachment to the pump, and the other serving as an air inlet (Figure 11-11). After collection of the air sample, the container is removed from the pump, and the water is poured through a membrane filter. The outlet of the membrane filter is attached to a vacuum flask for filtration. The membrane filter is then removed and transferred aseptically to a suitable agar medium. After incubation, the membrane is inspected for viable colonies. Further details on the membrane filtration portion of this procedure may be found in Gelman Sciences' *Laboratory Filtration* (1993).

DETECTION AND ELIMINATION OF
SYSTEMIC CONTAMINANTS—INDEXING

The detection of pathogens in culture material is affected by the growing environment. Cool and dry conditions will reduce pathogen activity, thus masking their presence and causing growers to unknowingly propagate symptomless but infected stock. Plant disease is often

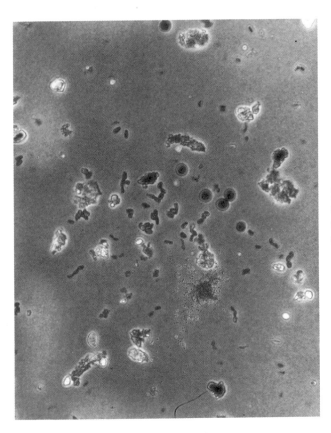

Figure 11-10a. Direct microscopic examination (×400) of dust on the surface of a membrane filter, showing bacteria, fungal spores, and inanimate matter.

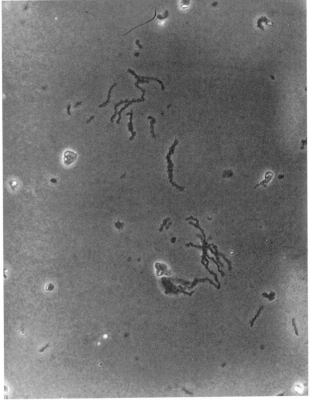

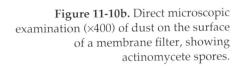

Figure 11-10b. Direct microscopic examination (×400) of dust on the surface of a membrane filter, showing actinomycete spores.

Figure 11-11. Pump with impinger (container) used for collecting and ascertaining viable particles present in air samples. Courtesy Mine Safety Appliances Co.

an economically limiting factor in the production of ornamental plants by cuttings. Systemic diseases are caused by various plant pathogens, such as the bacteria *Erwinia chrysanthemi* and *Xanthomonas dieffenbachiae*, which are highly pathogenic to several foliage plants (Knauss 1976).

   Of great concern are bacteria that attack vegetable crops, especially those crops that are propagated vegetatively, such as potatoes *(Solanum tuberosum)*. When infected, crop yield can be reduced by as much as 50% or more. Particularly offensive are *Erwinia carotovora*, which causes potato black leg (soft rot), and *Corynebacterium sepedonicum*, which causes bacterial ring rot. Some pathogenic fungi include *Verticillium albo-atrum* in chrysanthemums *(Dendranthema)* and *Fusarium oxysporum* in cuttings of carnations *(Dianthus)*, among others. A wide range of viruses also attack a great many plants of nutritious and ornamental value.

Because of such concerns, most states involved in potato production have programs for producing certified "seed" potatoes (vegetatively produced potato tubers that are used in place of true seeds). In these programs, seed potatoes are certified as being free of soft- and ring-rot bacterial infections, as well as having a low level (3% or less) of certain viral infections (leaf roll, mosaics, and spindle tuber viruses). After replanting 3 to 5 successive crops, the incidence of infection will increase, thereby decreasing yield to the point where new certified seed potato stock becomes necessary. New seed potato stock initiation and multiplication is done via test tube culture.

Certification programs have also been undertaken for several major fruit crops. For example, some states provide disease-free foundation stock of strawberry (*Fragaria*) plants. Usually these have been tissue cultured from heat-treated, virus-tested stock plants.

Systemic contaminants can be detected by various indexing methods. *Indexing* refers to any of several procedures for determining the presence of systemic, microbial, and viral contaminants in suspect plants. Many indexing methods are now available for detecting systemic pathogens in a variety of plants.

One respected but laborious method for detecting microbial contaminants in *Dieffenbachia* explants has been proposed by J. F. Knauss (1976). *Dieffenbachia* is frequently tissue cultured and is an important segment of the ornamental tropical foliage plant industry (Oglesby 1994); for that reason it is used here as the example for indexing. Because of the potential for severe losses from systemic pathogens, it is wise to index *Dieffenbachia* for *Erwinia chrysanthemi* and the virus dasheen mosaic. These pathogens are present in most stocks of certain *Dieffenbachia* species, thus making the production of these species without indexing a risky business (Taylor and Knauss 1978).

With Knauss's indexing method, explants of the suspect plant are prepared from lateral buds and apical meristems. When the explants reach a length of approximately 2 in (5 cm), tissue samples (slices) are removed and transferred to 4 different indexing media capable of enriching microbial growth. More slices are removed and transferred at 2 later time intervals. The time intervals must be long enough to enable sufficient new plant growth. Previous studies have shown that the likelihood of detecting latent (dormant) contaminants is improved by sampling at 3 successive time intervals, or index stages. The explant is discarded if microbial growth occurs in any of the 4 indexing media.

Indexing media

> Medium 1: Yeast/dextrose broth. pH 6.2. Suitable for growth of most major fungal and bacterial pathogens inhabiting tropical foliage plants.
>> Yeast extract   10 g/liter
>> Dextrose   10 g/liter

> Medium 2: Sabouraud's dextrose broth (commercial dehydrated concentrate). pH 5.6. General medium for fungus growth.
>> Peptone   10 g/liter
>> Dextrose   20 g/liter

> Medium 3: AC broth (commercial dehydrated concentrate). pH 7.4. This broth enriches growth of both aerobic and anaerobic bacteria.

> Medium 4: Trypticase soy broth (commercial dehydrated concentrate). pH 7.3. Enriches growth of both fastidious (exacting, a term used to describe microbes that require growth factors) and non-fastidious bacteria.

Equipment and supplies

Test tubes containing 5-ml aliquots of the 4 indexing media; 12 tubes of each medium are required per explant

Test tube racks

Labeler or marking pen

Forceps

Scalpel or knife blade

Sterile paper towel

Explant material

Test tubes containing modified MS media for explant initiation and rooting (see Section II, *Dieffenbachia*)

Paper bags, for covering test tube racks

Household tape

Procedure

Prepare and grow lateral bud and apical meristem explants according to the instructions in Chapter 8 and Section II, *Dieffenbachia*.

Stage 1

1. Label the test tubes of broth "1" through "4" to indicate the 4 media types and arrange them in the test tube rack. You will also need to indicate by number which tissue slices are in each of the 4 tubes of media; the 4 slices in tube 1 will be 1, 5, 9, and 13 (see Figure 11-12).

2. Aseptically place a piece of sterile paper towel on the work surface of the transfer hood.

3. Using forceps, aseptically remove the first explant from the agar growth medium and place it on the paper towel.

4. Using a sterile scalpel, excise the shoot from the root and discard the root.

5. Using a sterile scalpel, cut 4 cross sections (each section approximately 0.5 to 1.0 mm thick) progressively up toward the shoot tip—but always make certain to leave enough stem (approximately 25 to 50 mm) to enable rooting of the remaining stem tip. Ideally there should be sufficient stem material (8 to 16 mm) to prepare a total of 16 cross sections.

6. The cross sections are handled in groups of 4. From each group, place one cross section in a test tube of each of the 4 indexing media.

7. Repeat steps 5 and 6 with the next 4 cross sections. Hopefully, each of the four indexing media test tubes will contain 3 or 4 cross sections when you have finished. Place the test tubes in a test tube rack.

8. Transfer the remainder of the stem tip, stem base down, to a test tube of growth and root-development agar medium, and label the test tube plantlet #1. Place this test tube in a separate test tube rack.

9. Repeat steps 2–8 for each of the remaining explants. When finished, cover with a paper bag the test tube rack that contains tubes with explant cross sections. Seal the bag with household tape and incubate it at about room temperature (75 to 85°F [24°C to 29°C]) for 3 weeks.

10. Place the racks of test tubes containing plantlet shoot tips under standard lighting in the growing room.

11. Once a week vigorously shake the paper bags containing the test tubes with

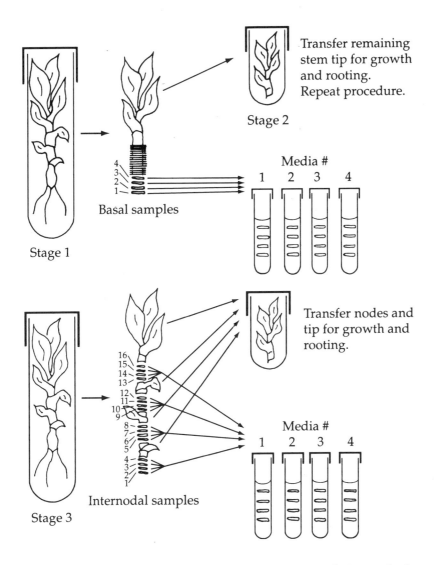

Figure 11-12. Illustration of stages 1, 2, and 3 of Knauss's indexing method for detecting systemic microbial contaminants.

the indexing media to introduce oxygen into the liquid media. Be careful not to wet the tube caps.

12. After 3 weeks, shake the indexing test tubes and examine for visual evidence of contamination—either clouding of the broth or, sometimes, visible growth. If either occurs in any of the indexing tubes, the plantlet line in question should be destroyed.

Stage 2

Repeat stage 1 indexing procedures and observations for the remaining explants that do not display evidence of contamination in stage 1. Discard all contaminated plantlet lines after completing stage 2.

Stage 3

Proceed with stage 3 indexing procedures for all explants that receive negative indexing results in stage 2. For this stage, the cross sections should be taken from the internodal area of the shoots.

1. Take the 4 groups of 4 tissue slices from internodal areas (see Figure 11-12), following the procedures described for stage 1. Repeat for all remaining explants.
2. Transfer the remaining nodes and shoot tips into test tubes of growth and rooting agar medium. Place the test tubes in a test tube rack and incubate in the growing room.
3. Proceed with the indexing procedures as described in stage 1. Following incubation, examine the indexing tubes for growth. Destroy any explant tissue showing positive indexing results.

The explant tissues remaining after the third stage of indexing are presumably free of microbial contaminants. These procedures for plantlet initiation and indexing *Dieffenbachia* are useful for eliminating pathogens from most other ornamental tropical foliage plants, although other media modifications may apply.

VIRUS DETECTION METHODS

Early knowledge of viral plant diseases dates back to written records of several hundred years ago. Notable examples are the potato leaf roll and tulip break, a color variegation found in infected tulips *(Tulipa)*. While most viral infections are detrimental, tulip break is considered beneficial because of the ornamental colorful streaks. Most plants infected with viruses eventually die, but occasionally, infected plants recover even if the virus is still present. The reasons for such recovery are not completely understood.

Many plant viral infections can be recognized by visible symptoms and further diagnosed by inoculation of an indicator plant with an extract from the infected plant tissue. Unfortunately for detection purposes, latency (the non-expression of visible virus symptoms) is often a feature of virus infection in plants growing *in vivo*, and it appears to be the norm for plants growing *in vitro* (Cassells 1992). Some of the more common visible viral lesions are mosaic and mottle patterns on leaves, chlorotic leaf spots, and yellow coloring. Other aberrations include dwarfism, ring spots on leaves, leaf curl, and leaf wilt (Hill 1984). Another common symptom of a virus is the formation of inclusion bodies in plant tissue. Their presence can be determined by preparing stained slides of the plant material and observing them with a microscope.

Viruses are obligate parasites consisting primarily of DNA or RNA. They depend on the metabolism of the host cell for their reproduction. Unlike bacteria and fungi, viruses cannot reproduce outside of a cell or in a nutrient medium. Once inside the host tissue, they reproduce from generation to generation within the host, behaving as other protoplasmic components of the host cells. While in this state, the virus can remain latent and escape detection. A similar phenomenon occurs in bacteria infected with a bacterial virus (bacteriophage), whereby the virus becomes integrated into the bacterial chromosome and replicates only when the bacterial chromosome replicates. A repressor protein produced by the bacteriophage prevents it from being released from the bacterial DNA. When the repressor protein is no longer synthesized or is inactivated, the bacteriophage is released from the host DNA. Once released, large quantities of viral DNA are rapidly synthesized, eventually bursting the bacterial cell.

A few bacterial viruses, the filamentous phages, are released from the host cell by a process called extrusion, which does not destroy the host cell. In this instance, the phage particles do not become integrated in the bacterial chromosome but remain in the cell cytoplasm. There is also a slow, continual release of virus from the host cell (Nester et al. 1995).

Plant viruses may be filamentous, rod shaped, spheres, or brick shaped (Figure 11-13). Some viruses that attack bacteria look like 4-legged spiders out of science fiction. They are so small that an electron microscope is necessary to view them. Occasionally a plant virus investigator will have an outside laboratory conduct a confirmatory electron microscope test, but most detection of plant viruses employs serological tests. A serological test is the study of serum reactions that occur between certain proteins (antigens, in this case the virus) and antibodies (serum globulins). Antibodies are produced by inoculating an animal with an antigen preparation. When sap from a plant suspected of being virus-infected is mixed with serum containing a specific antibody, an antigen-antibody reaction occurs. By using specific indicator chemicals, the antigen-antibody reaction can be detected colorimetrically (by matching colors).

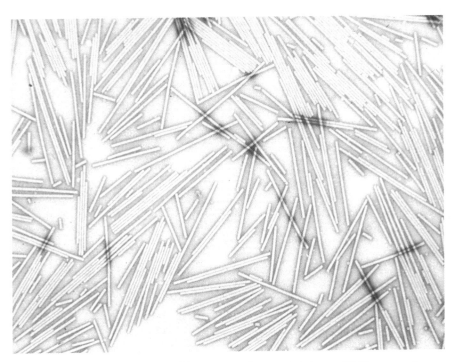

Figure 11-13. Tobacco mosaic, a filamentous virus containing an RNA core and a protein coat. This virus infects other plants besides tobacco *(Nicotiana)*.

ELISA and DAS ELISA testing techniques

For routine detection and identification of viral contaminants, an ELISA (enzyme-linked immunosorbent assay) procedure is commonly used. Until the development of ELISA, routine detection of viruses was impossible other than by employment of a large staff and the use of many indicator plants. Routine ELISA testing can reassure both grower and customer that plants are free from viruses. At least 6 companies worldwide offer reagents and accessories for ELISA detection of about 100 different plant viruses (Gugerli 1992). The ELISA tests are simple to use, reliable, and relatively inexpensive. They are sold as individual reagents or as kits, which include all the necessary reagents including positive and negative control samples.

When using prepared kits, determination of the optimal dilution is not required except for the test sample of suspect virus antigens. If you prepare your own coating antibody and related enzyme-labeled antibody, you must determine the optimum working dilutions. Working dilutions for the coating antibody and enzyme-labeled antibody that give the greatest amount of color in conjunction with the least amount of color in the control wells must be chosen for use with the double antibody sandwich (DAS) technique; a scheme for this purpose is shown in Figure 11-14. For many plant viruses, a coating antibody suspension of 1 mg per ml of water may be further diluted to 1 to 2 micrograms (µg/ml). Working concentrations greater than 10 µg/ml can reduce the intensity of the virus-binding reaction. Optimum enzyme-labeled antibody dilutions range between 1/5000 and 1/10,000 (Hill 1984).

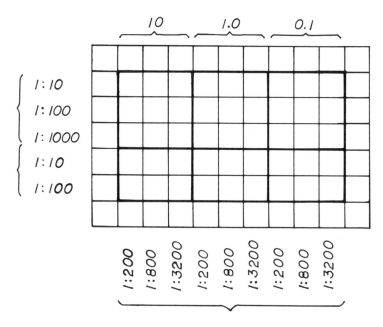

Figure 11-14. Scheme for preparing working dilutions of antigen and antibody used in the DAS ELISA test. Each square represents a reaction well with a particular dilution ratio.

The double antibody sandwich (DAS) ELISA technique is the most common among several modifications of the ELISA technique that have been developed (Hill 1984). The ELISA test is conducted on a polystyrene microplate that has a number of flat-bottomed wells for reactions; each well has a capacity of approximately 400 microliters (µl). The 4 major reactions are illustrated in Figure 11-15.

Reaction 1: Applying the coating antibody (Figure 11-15a)
1. In each reaction well, add the specific virus antibody desired. It will become adsorbed on the surface of the plastic. (Agdia, Inc., sells microplates that are already prepared, whereas some laboratories prepare their own antiserum, which they add to the plates after preparing the proper working dilution [Kaniewski and Thomas 1988].)
2. After adsorption (2 to 4 hours), wash the microplate with a buffer solution to remove any excess serum. Shake microplate until dry. The phosphate buffer solution (0.03 M, pH 7) contains 2% polyvinylpyrrolidone (PVP), 0.2% ovalbumin (prepared on the day of use), and 0.5 ml Tween 20.

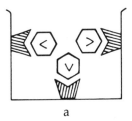
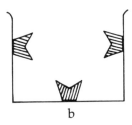

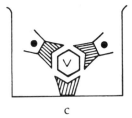
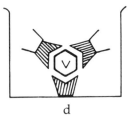

Figure 11-15. Four major reactions of the DAS ELISA test: a. addition of coating antibody; b. addition of leaf extract (virus); c. addition of enzyme-labeled antibody; d. addition of enzyme substrate.

Reaction 2: Addition of the antigen (Figure 11-15b)

1. Prepare a sap extract from a virus-suspect plant by mixing equal volumes of plant leaves and phosphate buffer, plus 0.5 ml of Tween 20 per liter.
2. Pulverize the mixture with a mortar and pestle or use a hand-held homogenizer, available with easy-to-roll bearings that glide over the surface of plastic bags containing the leaf/buffer sample (Figure 11-16).
3. Prepare a working dilution of buffered antigen (see Figure 11-14).
4. Add 200 µl of diluted antigen to several wells of the microplate, where it will become trapped and adsorbed on the surface of the specific antibody film.
5. For controls, fill wells with a sample of the buffer used for extraction and a positive control and a negative control. A positive control is an antigen that is able to attach to the surface of the coating antibody. It can attach because it has a physical configuration complementary to the physical configuration of the coating antibody. A negative control is an antigen with a physical configuration that is not complementary to that of the coating antibody.
6. After adsorption (2 to 4 hours), wash the microplate with the buffer solution to remove any excess antigen. Shake microplate until dry.

Reaction 3: Addition of the enzyme-labeled antibody (Figure 11-15c)

1. Add to each well 200 µl of a working dilution of the specific antibody complexed (labeled) with the enzyme alkaline phosphate. In the wells, the antibody will attach to another available reaction site on the antigen molecule.
2. After incubation (2 to 4 hours), wash the microplate with the buffer solution to remove any excess enzyme-labeled specific antibody. Shake microplate until dry.

Figure 11-16. Preparation of leaf/buffer sample.

Reaction 4: Addition of the enzyme substrate (Figure 11-15d)
> For this last reaction add 300 µl of freshly prepared substrate (*para*-nitrophenyl phosphate) to each well. If the antigen containing a bound antibody-alkaline phosphatase complex is present, a reaction will occur between the enzyme and the *para*-nitrophenyl phosphate to form a colored product. The time required for development of the color complex is usually less than 30 minutes. If a series of microplates need to be examined, the color reactions can be stabilized by adding 50 ml of 3 *M* NaOH to each well, followed by slight agitation of the microplate. Cover the microplates with a clinging plastic wrap and store in the refrigerator for several hours. The results can be determined either by visual examination or by inserting the microplate in a plate reader for colorimetric determination. Virus quantification is also possible by measuring light absorption with a colorimeter.

Confirmatory test using indicator plants for virus detection
1. Prepare sap extract (as described in reaction 2 above) from a plant that tested positive with the ELISA test.
2. Dust carborundum (with a grit particle size of 600 mesh) on one or more leaves of a known indicator plant. With your finger, gently rub the sap extract into the dusted leaves. The purpose of the carborundum is to lightly scarify the leaves for better penetration of the extract.
3. Store the plants in an isolated area and examine in 1 to 3 weeks for virus symptoms.

Chapter 12: Business

This book is devoted to tissue culture and not to business management, but certain important economic considerations must be addressed. As with any business, it is desirable to predict the financial success of a commercial laboratory, record it in progress, and periodically evaluate it in retrospect. Most operations cannot afford to operate without a tentative plan and a projected budget. Good management involves constant planning, study, observation, review, and adjustment.

MARKETING

Good marketing practices are essential for success. Marketing is an art and a science that must interact with management, technology, and common sense. It must equate with salesmanship, for without sales there is no commercial enterprise. If you have a superior product, you can probably find a market for it; but if you first find a need and then grow to fill that need, you will have an even better chance of selling your product.

Perhaps the most appropriate circumstance for establishing a tissue culture laboratory is as an adjunct to an existing nursery, which then becomes a built-in customer for the tissue culture operation. If the product does not sell as a microcutting, rooted plantlet, or liner (small, established plant), then there is room to grow it on within the nursery until the market is ready.

Sometimes an independent laboratory will form an association with a nursery that will serve as the desired cushion between production and market. The nursery can grow-on the plantlets on a consignment basis, or simply be an ongoing source of orders. An independent laboratory can act as a custom grower for various nurseries, or if a laboratory is tied to a nursery, it can custom grow for outside nurseries in addition to supplying in-house needs. Custom orders are a more comfortable approach than speculative growing, which has much uncertainty and risk.

Orders for woody plants or bulbs generally need to be placed a year in advance. Foliage plants require less time. If the laboratory anticipates certain orders, an inventory of stock cultures can be maintained to help cut down the lead time. As stated previously, stocks can be maintained on a minimal medium and refrigerated for slower growth to lengthen the time required between transfers.

It is important to have money up front for custom orders. Unless you have previous experience with a plant in culture or it is commonly micropropagated, you should charge a fee for establishment—US $500 to $1000 is not an unusual amount. If you are unable to find an established procedure for tissue culturing a desired plant, the labor required for getting it started and determining optimum nutrients and conditions can be quite costly. The agreement with the customer should state that all or part of the fee is at risk because you may not succeed in attempting to establish the plant in culture. Several laboratories might be growing for a single buyer if the customer has been shopping around; without any financial agreement you could be left with a crop and no market.

Certain large operators offer a brokerage service. With this service, brokers will buy and consolidate material from several small micropropagation facilities to fill large orders or special needs. In addition to the large brokerage companies, such as Vaughans and Ball Seed Co., individual brokers are constantly on the road to accomplish the same end. Brokers will buy plants from small producers at about 20% of the price. This may not offset production cost, but it is better than destroying unsold products. Occasionally larger nurseries resort to dumping, the practice of unloading an over-stocked crop at less than cost. Such action may be deemed necessary by the nursery, but it promotes an unstable market and places the smaller producers at risk.

Sometimes other laboratories will order a few plants *in vitro*, which is a way for them to bulk up stock and avoid development time and expense. Fulfilling such small orders for other laboratories can be a risk because those laboratories well might become your competitor. For this reason many laboratories will not sell axenic (non-contaminated) or *in vitro* plants. An alternative is to sell culture test tubes for their real worth, the cost of development—asking US $500 or $1000 per test tube will quickly discourage any such attempt to undermine your market.

A catalog is an excellent way of announcing the products or services you have to offer. Computers have brought attractive and affordable design capabilities to your desk. The catalog or flyer need not be elaborate, but it should be neat, easy to read, inviting, and factual. Provide information on common and scientific names, sizes, availability, prices, quantity discounts, and shipping charges. Be sure to include an order blank, full address, phone and fax numbers, and year of catalog issue. Other desirable but optional details are illustrations, directions for planting and care, other relevant information on plants, or company history. Advertising in newspapers or magazines is another useful way of getting your products and services known in the general market. Once people are aware of your business, your customers will be your best advertisements through word-of-mouth.

It is good business practice to study the markets. Find out what other laboratories are doing—if they will tell you. When a laboratory has invested time and money into research and operation, it is easy to understand why they might withhold information and hope to profit by their advantage. The other side of the argument is that the free exchange of knowledge pushes technology forward and more people will benefit in the long run. At the very least, by communicating with other laboratories and consulting their catalogs you can find out what they are growing and how much they are charging. Visit nurseries and wholesale florists to learn what is in demand. Read journals such as *Greenhouse Manager*, *American Nurseryman*, *Pacific Coast Nurseryman*, *Flower and Garden*, and *Horticulture* to learn what is being sold.

As you study the market, try to find out if there will be more profit in selling plantlets, liners, or gallons. What are the going prices? How much would each of these end products cost you and return in profit? Is it really cheaper to tissue culture the varieties in question than it is to maintain stock plants and grow from cuttings? How long will it take to establish a plant in culture? Boston ferns (*Nephrolepis*) may take 1 month, strawberries (*Fragaria*) 2 months, and *Rhododendron* 6 months. What is the rate of multiplication? How long is the rooting period? Is it cheaper to build a laboratory and produce the plantlets, or to have them custom micropropagated in an existing laboratory? Answers to these questions will help to answer the main underlying question: Will there be a profit?

A new helpful source of information and contacts went on-line in November 1994, a computer network whereby tissue culturists worldwide can readily communicate with one another.

WORLDWIDE MICROPLANT PRODUCTION

During the late 1970s and early 1980s there was a surge of new laboratories built in the United States. By 1990 competition, mergers, and buy-outs had succeeded in closing a number of doors. A 1986 survey indicated over 250 laboratories operating in the United States (Jones 1986). Surprisingly, according to a survey presented in newsletters and journals, there were only 125 in 1993 (Bridgen 1992). An accurate number of laboratories is difficult to determine. Some are minor adjuncts to various nurseries large and small. Others are large enterprises with or without nursery facilities. There are mergers, failures, and takeovers, and those institutions in support of research.

A limited but intensive survey of 46 members of the Tissue Culture Association was conducted by Dennis Yanny (1988) for his Bachelor of Science thesis at Cardinal Stritch College in Wisconsin. The 30 members who responded (10 were no longer in business and 6 did not reply) reported 701 full-time employees and 412 part-time employees, none of whom had a college education. The average annual production was 1.45 million plants, mostly foliage plants. Of these, 24% sold as plantlets, 44% as liners, and 32% as finished plants. The average initial investment was US $36,700 (minimum was US $10,000). The average current investment (at the time of the questionnaire) was US $236,000 (minimum US $20,000). Laboratory space averaged 2700 square feet (243 square meters); the smallest measured 1200 square feet (108 square meters).

One study has placed total micropropagated plant production in North America at 84.7 million per year (Jones and Sluis 1991). No production number was reported for orchids, but foliage plants dominated with 32.5 million per year. *Syngonium* had an output of 16 million plants per year, woody ornamentals were reported in excess of 13 million per year, and ferns were estimated at 12 million. *Gerbera* may be losing ground due to some seed now coming true to color. Other major micropropagated plants included bulbs, fruits, vegetables, and trees.

With increasing worldwide trade the tissue culturist will be competing with markets both at home and abroad. Western Europe and Israel combined accounted for an annual total of 212.5 million micropropagated plants in 1988. The major products reported were 92 million pot plants, 38 million plants for cut flowers, and 19.4 million fruit trees. The Netherlands alone produced 16 million *Gerbera*, 13 million bulbs/corms, 12.8 million ferns, 2.4 million orchids, and 4 million African violets *(Saintpaulia)* (Debergh and Zimmerman 1991).

Eastern Europe is credited with producing 55 million micropropagated plants per year. A limited report from Asia estimates annual production at 74 million, which includes 44 million orchids. Production numbers from South and Central America are also difficult to find. Orchids, bananas *(Musa)*, *Eucalyptus*, carnations *(Dianthus)*, *Anthurium*, and potatoes *(Solanum tuberosum)* are the most common products of South and Central American tissue culture operations. One Venezuelan laboratory produces 1.5 million bananas per year. Statistics on micropropagated products from Australia and New Zealand are incomplete, but 20 laboratories reported a total of 25 million plants. Except for a few ornamentals, grown primarily in Egypt, most plant tissue culture in Africa is dedicated to breeding and improving food crops (Debergh and Zimmerman 1991).

With this review of worldwide activity in the field, a person contemplating starting a tissue culture business may be discouraged. One might conclude that there is a minimum of profit to be had from standard tissue cultured products, such as foliage houseplants, rhododendrons, and some perennials. If that is true, then the greatest profit lies in being the first with plants newly issued, patented plants (especially if it is your own patent), plants that are difficult to propagate, and unique plants.

Many foreign markets are not satisfied by internal production. The best way to find these markets is to travel to the countries that have potential buyers. There are special hurdles in foreign markets, primary among these being customs. There was a time when plants *in vitro* were considered free from disease as long as no contaminants were visible. It has become common knowledge, however, that *in vitro* plants may have endogenous pathogens, the effects of which will not be visible until long after the plants have left the culture jars. Consequently, customs requirements vary with time, country, state, and the type of plant. Often a phytosanitary certificate is required, which involves a government inspection and an inspection fee before material can be shipped. Where there are regulations, the minimum requirement is that the plants have never been in soil (as opposed to soil mix or artificial soil) or, sometimes, that the soil be washed off. Everyone loses when shipments are delayed or destroyed, so check with appropriate government agents to avoid potential problems.

COSTS

Because a successful operation usually will expand, the original planning of the physical structure of the facility should provide for potential growth; otherwise the result will be a piecemeal, expensive, disorganized assortment of additions. A local contractor can help determine facility cost, but shopping for yourself with a detailed plan in hand can save considerable expense.

Equipment can be changed more readily than a building. Scientific equipment is fine, but oftentimes cheaper household equivalents can simply be adapted; for example, a kitchen dishwasher instead of a laboratory glasswasher, or a pressure cooker instead of an autoclave. Alternatives should be studied and balanced against the size and nature of the operation being considered. In other words, do not build a lab for 150,000-plant annual production and then buy equipment for a million plants, nor should you build a lab for a million plants and try to get by with household equipment.

Detailed studies of tissue culture costs are available in the literature to determine profit or loss (Anderson and Meagher 1977; Anderson and Meagher 1978; de Fossard 1993; Standaert-de Metsenaere 1991). Records should be kept of overhead costs, purchases, labor costs, and losses. Ideally, you should itemize overhead, investment, depreciation, chemicals, transfer rate, multiplication rate, growing-on costs, labor, losses, and shipping costs. Then determine a profitable price. If you have not yet tissue cultured anything, you will have to base your figures on a survey of the market and your best educated guess. If you are operating an efficient laboratory and can afford cost analysis to determine your pricing, your prices will probably be in line with demand-oriented or competition-oriented pricing. If you omit the analysis and simply use competition-oriented pricing, you will quickly know whether or not you are making any money. If you are selling at a loss, an analysis might help you find where you can save money.

Labor is the greatest cost in a laboratory. Labor costs—which involve media making, transferring, culture surveillance, and benefits—can account for as much as 80% of operating costs (Jones and Sluis 1991), so management must constantly watch for ways to maximize labor efficiency. The relative cost of premixed or partially premixed media should be studied and weighed against the cost of labor to mix it; however, the desirability, or necessity, of having the flexibility of varying the ingredients in media should not be underrated.

The preliminary cost projection should stipulate the number of employees required and their qualifications. Personnel should already have some knowledge of tissue culture, sterile technique, chemistry, and plants, or it will have to be taught. Common sense, cleanliness, a sense of humor, and the ability to get along with others is of more importance than any technical background, which can be learned on the job.

Regardless of background, an employee must have a clear understanding of the difference between research and production. The inquisitive will want to experiment, the overly precise will want to pick at transfer material or dawdle with labeling or records. There is no faster way to lose profit. Problem material will, of course, require special treatment and will slow up production, and these matters should not be left to the amateur. Efficient mainstream production is at the heart of successful nursery tissue culture.

Most larger plant tissue culture laboratories have a research and development (R & D) department that may be responsible for studying problems in production, new plant varieties, sanitation, safety, and quality control. Production problems include media formulation, multiplication rate, root initiation, and detection and elimination of microbial and viral contaminants. Other functions of the R & D group may be maintaining cultures in storage and training new staff.

Genetic engineering is another function of R & D. The department may be assigned to develop new hybrids using transformation (introduction of foreign genes into a plant), or by protoplast fusion. They may be asked to fingerprint plant material (identify DNA or proteins using electrophoresis) or to investigate the production of secondary products *in vitro*. See Chapter 13 for a discussion of these and other biotechnological processes.

If no one within an existing nursery organization is qualified or available for training to direct the tissue culture operation, advertising in technical and trade journals or the local newspaper frequently can turn up the right person for the job. College placement offices can assist in identifying technicians. Oftentimes competent personnel can be located through meetings of the International Plant Propagators' Society, the International Association for Plant Tissue Culture, or the Tissue Culture Association (see Appendix C).

The periods during which transfer activity is reduced is an ideal opportunity for the established grower to rotate personnel from laboratory to nursery. Cultures targeted for spring outplant make winter the busiest time of the year inside, but it is usually a slow time outside in the nursery. The advantages of retaining year-round employees are evident to employers who must hire, train, and evaluate new employees.

Labor-saving devices are generally a good investment. Usually the more automated an operation is, the lower the labor costs; on the other hand, more automated equipment is also more expensive, such as using an autoclave instead of a pressure cooker, an automatic media dispenser instead of a pitcher, an electronic balance, or an automatic labeler. If capital is available for labor-saving equipment, the investment will pay for itself in some defined period of time.

Travel time, whether from one room to another or the distance and frequency the hands must travel in operating equipment or making a transfer, also translates to dollars. Passthrough windows in a small operation save time by eliminating the need to carry or cart cultures and materials through doorways from room to room. Similarly, careful organization of equipment around the transfer technician or in the media preparation room contributes to the most efficient use of labor.

The size and shape of culture containers is another economic concern. Test tubes are valuable for starting cultures and restricting contaminants; larger containers are more economical in terms of transfer time. The adept technician can transfer approximately 130 tubes per hour when dealing with good material and a 3- to 6-times multiplication rate. However, transferring cultures into pint jars or Magentas at a rate of 20 containers per hour, each with 16 propagules, is equivalent to 320 test tubes per hour.

Losses from contamination of tissue cultures, especially from disease in Stage IV, can be devastating. The sooner contamination is detected and disposed of and/or treated and appropriate prevention measures are taken, the more cultures and plants will be saved. Contamination can wreak havoc on a good plan. A realistic plan will take into account some overage for losses, which should not exceed 2%. Any greater amount calls for serious investigation. Money spent for air filtering, sterile transfer, and containers that prevent contamination is money well spent, as is time spent cleaning surfaces.

The fear of contamination is well founded, but operations can spend initially an excessive amount attempting to meet hospital-quality sterility. Purchasing, replacing, and cleaning special clothing, such as gowns, goggles, masks, gloves, and slippers, can be very expensive without being effective. Gloves are advisable primarily to protect the hands of workers from harsh disinfectants, and secondly to protect the cultures. The effectiveness of laboratory coats is questionable because they may be no cleaner than normal clean clothing. If a problem with contamination emerges, track down the source and then buy what is needed to combat it. Experience, coupled with a contaminant quality-control program, will demonstrate which sanitary practices are most effective for you.

The use of refrigeration for some cultures will slow down contamination as well as culture growth if the cultures cannot be cared for on time. Refrigerated storage of cultures is a tool that can be incorporated either deliberately in original plans or as a recourse to ease work loads when there is too much to transfer at one time. If cultures become aggressive and are ready for transfer ahead of schedule, it is possible to refrigerate some species until it is time to transfer them. Strawberries *(Fragaria)*, *Rhododendron*, carnations *(Dianthus)*, and chrysanthemums *(Dendranthema)* are some of the plants known to tolerate several weeks or months of storage at temperatures of 34 to 36°F (1 to 2°C), usually without any light.

Occasionally, natural lighting can be used in tissue culture growing rooms to conserve energy consumption. Because natural light fluctuates significantly and is difficult to control, however, it is not recommended, and power usually is not a major expense. Fluorescent lighting is constant, relatively inexpensive, and usually provides more than enough heat required in the growing room. It is economical to schedule the dark cycle during the day, which saves on cooling costs, and maintain the light period during the night to save on heating costs, especially if the heat from the lights is circulated throughout the laboratory, minimizing the need for other heat sources. Even when ballasts from fluorescent fixtures are relocated outside the growing room to prevent overheating and reduce cooling requirements, they still provide a heat source that can be tapped by the air circulation system. Turning off lights and appliances not in use and providing good insulation are basic to energy conservation.

Saving on media chemicals by buying inferior quality is not a wise practice because it will have a high impact on plant performance for a relatively small percentage savings. Most often the extra cost of buying quality chemicals and supplies will pay off in the long term.

A tissue culture operation offers savings in space compared to a conventional nursery operation. Less space is required to carry stock plants, to start plants, and to multiply them. With careful scheduling of laboratory and growing-on space, the grower need allocate fewer dollars to space and will realize a higher return on the dollars invested in land and buildings.

CROP PLANNING

A production plan for each of the crops being micropropagated must be a high priority (Table 12-1). A manager should develop a time frame, especially for crops that must multiply to numbers in the thousands by spring delivery. A good plan starts with the number of plants ordered and the scheduled delivery date. Working backward toward the explants, target numbers are established and the container requirements noted for specific dates.

After a crop plan has been established for each variety of plant, an overall space schedule must be made to assure adequate space and product flow. Usually, as numbers increase, cultures are relocated. When these moves are planned ahead of time, there will be a minimum of confusion and wasted motion. In addition, you should devise a plan for *in vitro* media and an estimate of the labor hours.

When you have about enough plantlets to fill your order, a table can help you determine

Table 12-1. Sample Crop Production Plan for 12,000 Field-Ready Strawberries
Strawberries, Kings 12,000; Code: KNGA; Order: #662; Ship 4/1/97

Date	Number	(out to)	Trays	6-paks	Jars	Tubes
4/1/97	12,000	Ship				
2/1/97	12,250	Greenhouse				
12/15/96	12,500	Acclim.(2)	174	72/tray		
12/1/96	12,800	Acclim.(1)	64			
11/1/96	12,800	Rooting, singles			512	
10/15/96	3,200	Multiply			67	
10/1/96	1,067	Multiply			23	
9/15/96	350	Multiply			7	
9/1/96	128	Multiply				128
8/15/96	60	Multiply				60
8/1/96	30	Multiply				30
7/15/96	20	Transfer				20
7/1/96	25	Explants				25

how many to put into rooting and how many into multiplying in the last 2 transfer cycles. For example, say you need 25,000 plants. At 30 plants per rooting jar, you will need 833 rooting jars for 25,000 plants. You have on hand 200 multiplying jars. The plants have been multiplying at the rate of 2 to 1. You judge that each multiplying jar will fill 3 rooting jars (clumps are usually singulated as they are transferred to rooting jars). Thus the plantlets in your 200 initial multiplying jars will require 400 multiplying jars or 600 rooting jars after the first transfer. This gives you too many multipliers (requiring 1200 rooting jars next transfer) but not enough rooters this transfer, so you need to determine how many should be turned out to rooting this transfer in order to have 833 rooting jars after the next transfer.

Make a chart with jars to go to multiplying on one axis and jars to go to rooting on the other—the sum where the 2 axes meet is the 200 jars you have on hand. The chart could run from 0 to 200 on both axes, but a likely range is illustrated in Table 12-2. The chart shows that if plants from 80 of the 200 jars are transferred to 160 multiplying jars, the next transfer will produce 480 rooting jars (i.e., the 80 multiply to 160, which are then transferred to 3 times as many rooting jars, 480). The remaining 120 of the original 200 multiplying jars will fill 360 rooting jars this transfer. With this combination—80 jars to multiplying and 120 to rooting—840 rooting jars will be produced by the end of the next transfer.

Planning may appear to be a futile exercise, but as you plan you will find it enlightening to make up a set of schedules. Before you finish, you will realize that you are able to visualize the total operation much more clearly.

KEEPING RECORDS

Most nurseries and tissue culture operations utilize one or more computers to help with record keeping. Adapt the programs of your choice for keeping records on production, performance, and finances. To avoid the need for everyone to use the company computer, technicians can orally record their hours, observations, production records (code, date, number of jars in and out), and other information on a tape, which later can be entered appropriately into the computer. Or they can keep careful written records to be later entered into the computer. The careful records that are required for research projects are usually not economically justifiable for basic production purposes.

Table 12-2. Final Transfers to Achieve Required Stage III Plantlets.
M = multiplying jars; R = rooting jars.

Jars to Rooting	Jars to Multiplying			
	60	70	80	90
140	120 M × 3 = 360 R + 420 R ——— 780 R			
130		140 M × 3 = 420 R + 390 R ——— 810 R		
120			160 M × 3 = 480 R + 360 R ——— 840 R	
110				180 M × 3 = 540 R + 330 R ——— 870 R

Records are useful only if they are referred to, and not just buried in a drawer or in a file on the computer. The object of record keeping in production is to provide sufficient cross checks for peace of mind, without creating costly busy-work. At one end of the spectrum of record keeping is the identification of each explant and its source plant and the maintenance of this identity throughout the history of the clone. Where a number of clones of the same cultivar are involved, such meticulous record keeping facilitates the elimination of a clone if contamination or other problems develop. At the other end of the spectrum is a running file of each crop, which provides the customer's name and address, the name and code of the cultivar, date started, transfer dates, number transferred, media used, location (on numbered growing room shelves), culture conditions, tests performed, and results. Also important is the almost indispensable log of telephone calls. A guest book, an address book, and a phone book may seem obvious, but such tools could be easily forgotten until it is too late.

Another invaluable record is a chemical inventory, noting when a new container of chemical is opened, when and where it was purchased, the cost, and where it is stored in the laboratory. A reliable system for reordering chemicals will avoid the problem of suddenly being without one when you need it.

A carefully established and maintained record of general financial accounts is essential. It is easy to lose money quickly on a tissue culture operation. It can be a year or more before any money comes in and several years before you know if you have been successful. At some point you must answer the question, Are we making any money? A good set of financial records, meticulously maintained, will provide answers to this question and is the key to successful financial management.

The foregoing suggestions, coupled with professional advice on business planning and projections, should provide a firm foundation on which to answer some of the questions posed earlier and to decide whether or not you want to pursue starting a tissue culture business.

SHIPPING

The best method of packing tissue culture materials for shipping depends on the stage at which the micropropagated plants are shipped. Rarely is it necessary or desirable to ship plantlets in a sterile condition. To ship plantlets in the glass containers in which they have been growing is costly and the potential for breakage is high. Shipping plantlets in autoclavable plastic containers or aluminum food trays involves less cost and risk. Preferably they should be shipped without agar because it is likely that a container will not remain upright throughout shipment. If the order is for microcuttings, they can be placed in a clean, firm plastic storage dish with a tight cover. If padded with damp paper toweling, the microcuttings will not have room to tumble around in the container. Sometimes they can be placed in plastic bags. In either case, the container should be placed in a box and surrounded with plastic (styrofoam) packing peanuts. If rooted plantlets are to be shipped, a good method is to jelly roll them in moistened (not dripping wet) paper towels. Remove the rooted plantlets from the jars or rooting trays, and roll about 50 of them in moistened paper towels. Pack 10 rolls in a plastic bag, and place 6 to 12 plastic bags in a small carton lined with styrofoam sheets to protect against excess heat or cold during shipment.

When shipping larger plants, ship them as you would any other nursery stock. They can be removed from the pots or plugs and laid down in rows in a carton that has been lined with plastic or has been waterproofed inside. If they do not fit tightly, use plastic peanuts to fill in air spaces. Do not use the biodegradable peanuts that melt in water because the moisture in the plants will destroy them. For easy dispensing, hang a bag of peanuts from a rafter and attach a 2-in (5-cm) diameter hose with a shut-off valve so that they can flow by gravity into the boxes and be shut off at will. Seal and ship the boxes by usual methods. Assuming moderate temperatures, the plants will not need air if shipments will be opened within 3 or 4 days. For any longer length of time the plants should be refrigerated.

Chapter 13: Biotechnology

GENETIC ENGINEERING

In a broad sense, both tissue culture and genetic engineering can be defined as biotechnology—given an understanding of biotechnology as the human manipulation of biological material—but the distinction between the two fields is as important as their parallels. Tissue culture or micropropagation does not involve the deliberate changing of the genetic material of an organism, but rather multiplying the genetic material. Genetic engineering can be defined (in part) as the purposeful altering of genetic material, and in this way tissue culture serves genetic engineering, readily multiplying the plant material that is needed for and produced by genetic engineering.

Humans have been tampering with genetic material for centuries. Selective harvesting, hand pollinating or cross pollinating of plants, and multiplying or inducing mutants are just some of the ways in which humans manipulate genes and other biological processes. Almost all commercially available foods and most breeds of domesticated animals are the result of human interference, and the advances in the field of molecular biology have helped to further improve and fine-tune genetic manipulations. Most genetically engineered medicines are accepted without question, as are most "natural" mutants such as the navel orange. But all too often, genetically engineered food products, such as the "Flavor-Saver" tomato, with its increased shelf life and superior quality, are met with protest and rejection.

"Fruit that Swim" was the title of an article that appeared in *Financial World* (June 7, 1994). Although the title may be humorous, the subject is serious, discussing the attempts to transfer into fruits the gene for an anti-freezing protein that is found in some fish. The article addressed the misunderstanding of genetic engineering that is held by a segment of the general public, that such tampering with nature can only be detrimental.

Expanding populations will require more food grown in less space with less water, with greater tolerance to weather and pests, and with better nutrients and keeping quality. Fortunately, industries and governments are working together to provide such food products, developed with sensible controls, and to educate the public regarding genetically engineered products. There is no way to halt progress, but it is possible to harness and guide it into safe, positive, constructive pathways. The field of biotechnology is undergoing an explosion of new research and new results, and the potential for further advances is remarkable.

Some purposeful applications of genetic engineering include providing a better understanding of cellular functions, learning how the genetic code directs the formation and maintenance of an organism, and improving cultivated plants by introducing traits of nutritional, esthetic, utilitarian, pharmaceutical, or horticultural value, such as insect and disease resistance (Greenberg and Glick 1993). Some of the plants, mostly food crops, that have been transformed by inserting foreign genetic material include soybean *(Glycine max)*, rice *(Oryza)*, corn *(Zea mays)*, potato *(Solanum tuberosum)*, lettuce *(Lactuca sativa)*, cotton *(Gossypium)*, and tomato *(Lycopersicon)*.

The expanding role of plant tissue culture as an accompanying tool in genetic engineering and other plant biotechnology warrants a discussion of the molecular basis of life. Even a superficial concept of what is happening in the world beyond our vision increases our understanding of new products in the markets, of research articles appearing in newspapers or journals, or of events in our own laboratories and greenhouses.

Though it certainly is incredible that, with nuclear material so infinitely small, scientists can cut out genes and move them into other organisms, any study of nature reveals that the combining of genes is nothing new. Consider, for example, a virus. These amazing bits of DNA or RNA reside in a protein coating. When separated from living material, viruses are virtually inert chemicals. But if a virus enters a living cell, whether bacterium, animal, or plant, the virus suddenly comes to life. Settled on the wall of an organism, it shoots its own DNA out of its coating through the organism's cell wall to merge with the organism's DNA, which then becomes *recombinant* DNA (DNA combined from different sources). In such an instance, the viral DNA replicates only when the host cell's DNA multiplies. Sometimes a virulent virus can infect the host cell, producing a diseased state in which the virus replicates itself until the cell wall bursts and a myriad of new viruses spread out to find other cells to victimize. The only defense an organism has is its restriction endonuclease enzymes, which act as scissors to cut out unwanted viral DNA. Scientists have isolated numerous restriction endonucleases, enzymes that cut different chromosomal DNA base sequences.

The major emphasis of the biotechnological revolution has been directed at zoological subjects rather than botanical ones. At the molecular level, however, there is very little difference, and therefore most of the information is applicable to both fields.

Various metaphors have been used to describe the double-helix structure of DNA (Figure 13-1), having been compared to a circular stairway, a ladder, or a double strand of beads. If you were to unwind a spool of thread and form many small loops, which you then wad up in your hand, you could begin to appreciate the mass and proportion of the DNA threads that make up the chromosomes within the nucleus of a cell. Picturing the thread as a long microscopic zipper can further define the complexity of a DNA molecule. Consider each of the interlocking pieces of the zipper as representing one of the 4 nuclear bases—adenine (A), thymine (T), guanine (G), or cytosine (C)—that code for all the processes of living organisms. A and T are always paired across the strand from each other, as are C and G. A base pair is held together with hydrogen bonds and, with its contingent sugar and phosphate, is called a *nucleotide*. On the outer edges of the long molecule is the sugar-phosphate backbone, the handrails of the metaphoric staircase. Considering that only 4 bases control all cells, one must wonder how characteristics can vary so widely from one species to another, or even within an individual. The reason is that the base pairs can align in various numbers and sequences.

DNA serves as a template for RNA. When the RNA polymerase enzyme "unzips" the double strands of DNA, a single-stranded complementary messenger RNA (mRNA) is formed according to the coded fragments of the DNA. This mRNA then flows from the nucleus into the cytoplasm of the cell to the ribosomes where the proteins are manufactured.

Over 90% of DNA is presumed non-functional because it does not "read" the same as that which is known to be functional. Intermixed in sequences of nucleotides are nucleotides that do not code for anything, as far as is currently known. These are called *introns*. The nucleotides that are known to be productive are called *exons*. After the DNA is *transcribed*, producing the single-stranded complementary RNA, the RNA removes the introns with restriction enzymes. The remaining functional sequences, the exons, are spliced together to form mRNA.

A *codon* is a group of 3 nucleotides that code for a particular amino acid (Figure 13-2). A *gene* is a section of DNA, made up of several hundred nucleotides, that codes for a protein. There are 20 kinds of amino acids that join in various numbers and sequences to form proteins. Hundreds of proteins can be found in a single cell (Tortora et al. 1982). Proteins serve as enzymes, organic catalysts that assist reactions but are not used up in the processes. Proteins also play a structural role in the make-up of cell components such as cell walls, membranes, and chromosomes.

When a cell is ready to divide (multiply), the folded mass of DNA further condenses into

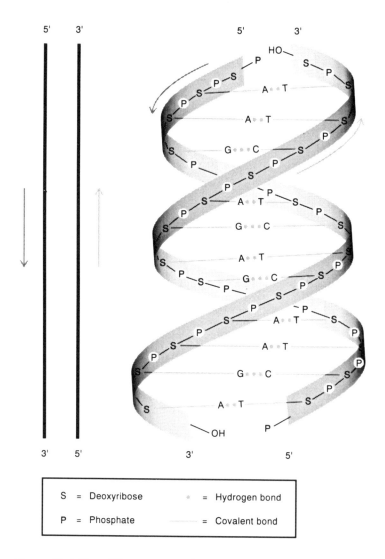

| S | = | Deoxyribose | • | = | Hydrogen bond |
| P | = | Phosphate | ——— | = | Covalent bond |

Figure 13-1. Two different ways of representing the DNA molecule. The 2 complementary strands run in opposite directions (antiparallel), with the 5′ end containing a phosphate bonded to the sugar and the 3′ end having a hydroxyl group (OH) on the sugar. The 2 dots between the A and T indicate that 2 hydrogen bonds exist between these molecules; the 3 dots between the G and C indicate that there are 3 hydrogen bonds. The purines (A and G) and pyrimidines (C and T) are on the inside of the molecule. From Nester et al. 1995, reprinted with permission from William C. Brown Publishers.

chromosomes. The DNA is "unzipped" by a polymerase enzyme that travels down the double strand, separating the 2 strands. As a double strand is separated, a new complementary strand forms, so that once again a double strand is formed. The new pairs move to opposite sides of the cell so that a new cell wall can form between them and divide the cell into 2 distinct cells. The new cells contain nuclear material that is identical to that of the previous cell and to that of all the other cells in that organism. In this vegetative process of cell multiplication, called mitosis, the diploid complement (2*n*) of chromosomes remains the same.

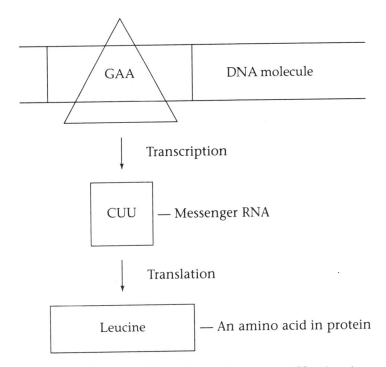

Figure 13-2. The codon GAA codes for the amino acid leucine via the messenger RNA CUU.

The sexual process of cell multiplication, called meiosis, is one in which the diploid complement of chromosomes is reduced from 2*n* to *n*; it is a reduction division. The products of meiotic division are the reproductive cells, the gametes or germ cells, also known as a microspore or pollen (male) and a megaspore or ovule (female). In the final step of the sexual process, the diploid chromosome number is reconstituted when the male and female gametes unite to form the fertilized egg, which then grows into an embryo. The random separation of chromosomes in meiosis accounts for the variability in offspring. Additional variation occurs when chromosomes are paired, at which time they may exchange genetic material; this is called *chromosomal crossover*.

That which some plants do naturally, genetic engineers do in a test tube. They integrate pieces of foreign DNA into pieces of plasmid DNA, which when inserted into a suitable host bacterial cell, replicate and produce useful products, such as insulin, interferon, and certain antibiotics. The plasmid DNA can also be inserted into plant cells. A *plasmid* is a small, extra-chromosomal, circular piece of DNA in bacterial cells that replicates independently of the chromosomal DNA.

The insertion of foreign DNA into a plasmid requires 2 different enzymes (Figure 13-3, steps 1 and 2). The first enzyme is a restriction endonuclease-type enzyme. Such enzymes recognize and become attracted to specific base sequences in double-stranded plasmid DNA, where they cut the DNA. The same restriction endonuclease is used to prepare the foreign DNA. The net result is overlapping cuts with sticky ends on both the plasmid and foreign DNA molecules, which enables the foreign DNA molecule to fit into the plasmid DNA. The second enzyme, ligase, sews these parts together, once again forming a circular plasmid molecule.

The next step in this process of genetic engineering is the introduction of the recombinant plasmid molecule into the host bacterium (Figure 13-3, step 3). This often occurs naturally if the

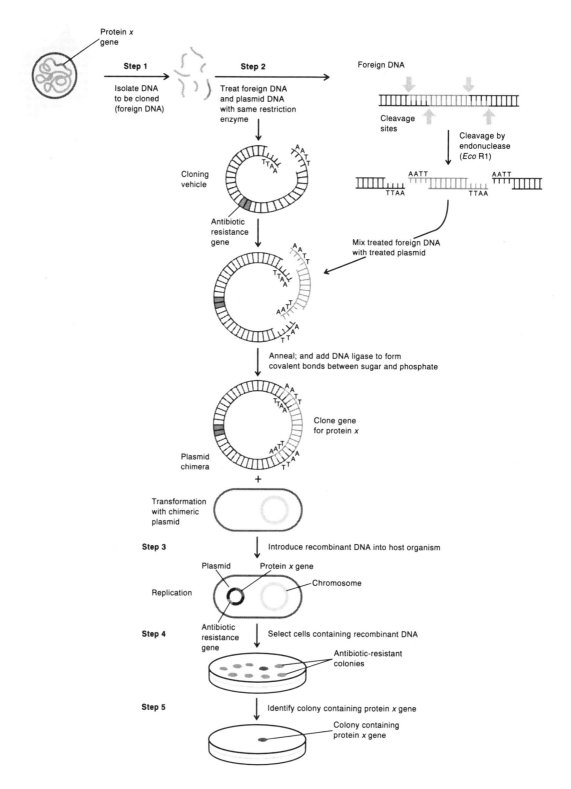

Figure 13-3. The scheme for forming and cloning recombinant DNA using a bacterial host cell. From Nester et al. 1995, reprinted with permission from William C. Brown Publishers.

host cell is competent. When not competent, special techniques are used to introduce the recombinant plasmid DNA. One of these techniques is *electroporation*, whereby the host bacterium and the recombinant DNA are mixed together and treated with an electrical current. The electrical current peppers the host cell wall, thereby enabling the fragmented DNA to enter the host cell and become incorporated into the host cell chromosome by homologous recombination (Figure 13-4).

In order to identify the recombinant DNA clone, the plasmid used for cloning the foreign DNA must contain a marker gene, for example a gene for resistance to a certain antibiotic. Such a marker enables selection of the host cells when they are placed on a growth medium containing the antibiotic (Figure 13-3, step 4). The next task is to identify the bacterial colonies that contain the foreign molecule. One way to do this is to use a probe. A *probe* is a single-stranded piece of DNA, labeled in some way (radioactivity, for example) for easy detection, that is complementary to some characteristic part of the DNA being tested, to which it will bind (Figure 13-3, step 5).

DNA probes have many applications. They can be used to detect *Aspergillus flavus*, a mold that produces aflatoxin, a hazardous food contaminant that causes billions of dollars in damage to corn crops. The DNA sequence of *A. flavus* has been determined and can be used as a probe. Under development is a sensitive probe test that can be used in the field, allowing the grower to know within hours if the mold is present. Previous cultural methods took at least a week for this determination (Cooke 1991).

Bacterial host cells capable of transferring their plasmid into a plant are frequently utilized in the genetic engineering of plants. The plant pathogen *Agrobacterium tumefaciens* has this

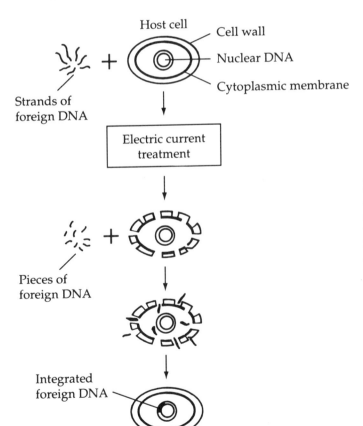

Figure 13-4. Electroporation. Application of an electrical current allows the donor DNA fragments to enter the host cell and become incorporated into the host cell chromosome(s), a form of genetic transformation.

capability. The plasmid of this bacterium contains tumor-producing genes that, when introduced into the plant, incite the formation of a tumor known as crown gall (Figure 13-5). It is now possible to remove these tumor-producing genes and substitute useful genes, such as genes for herbicide resistance (Gasser and Fraley 1992). With such plants, herbicides can be used to kill weeds without affecting the genetically engineered plants. Among other traits that are introduced in such a way into plants are insect and viral resistance, stress tolerance, altered fruit ripening, and altered flower color (Greenberg and Glick 1993).

Another method for plant gene transfer is the use of polyethylene glycol (PEG) for direct intake of DNA by plant protoplasts. Electroporation can also be used to move foreign DNA into protoplasts. The major drawback to using protoplasts is the long waiting period for the single cells to grow into plants of significant size.

Biolistic methods (particle guns) now allow direct transfer of DNA into any mature plant. Tungsten or gold microparticles are coated with the DNA to be injected, and they are then shot into plants at a speed of 1000 miles per hour (about 1600 kilometers per hour). The projectile of the "Bioblaster" is driven by an exploded .22-caliber blank cartridge. Thousands of cells are pierced and inoculated at once by this method (Glick and Thompson 1993; Wood 1989).

Figure 13-5. A natural form of genetic engineering. Production of a crown gall tumor on a sunflower (*Helianthus*) plant by *Agrobacterium tumefaciens*.

DNA fragments are commonly analyzed by electrophoresis (often called fingerprinting), a system whereby substances are separated according to the distance they travel on a gel within an electrical field—the smaller fragments will move faster and farther than the larger ones. After the fragments have separated, the current is shut off and the separated fragments are photographed, measured, compared with controls, identified, and sampled for further analysis, treatment, or transfer. Segments of the gel can be used as probes.

An important application of electrophoresis is the isolation of DNA, proteins, or enzymes that code for certain variables in different varieties of plants. A relatively simple electrophoretic test was reported for separating related enzymes (isozymes) that identify differences in red raspberry *(Rubus idaeus)* cultivars (Cousineau et al. 1993). Similar separations are made to verify and protect patented plants. A microcomputer-based image processing system can be used to classify isozyme patterns (Dixon et al. 1990).

Arabidopsis thaliana, mouse ear cress, a common weed in the mustard family (Cruciferae), is a favorite experimental plant for genetic engineering because it grows well in culture, it can flower in 5 weeks, produce thousands of progeny in 10 weeks, and has the smallest known genome of any higher plant—10 chromosomes (=2n) and 70 million base pairs. It has been used extensively to study mutagenesis and the selection of mutants.

The number of chromosomes varies greatly from one organism to another. Most flowering plants have between 14 and 22 (=2n) chromosomes, but sugar cane *(Saccharum officinarum)* has 80; American elm *(Ulmus americana)* has 50; tobacco *(Nicotiana)* has 48; and corn *(Zea mays)* has 20 chromosomes. Some ferns have as many 500 chromosomes. Humans have 46 chromosomes, a thread worm has 2, and one rhizopod (a protozoan) has 1500. The bacterium *Escherichia coli* has only one circular chromosome with 4 million base pairs (Tortora et al. 1982).

We must not lose sight of the fact that, in spite of certain genetic mandates, each plant is still an individual. Each is unique in some respects because of the interaction of its genes and their products, which are affected by the environment, and in turn, the influence of the environment on the interactions of genes and their products.

Two genetic engineering projects of worldwide interest and endeavor are the mapping of the human genome and mapping the genomes of major crop plants (Miksche 1994). The purpose of a gene map is to relate various traits to specific pieces or fragments of DNA. This knowledge has tremendous potential for assisting the cure of certain diseases and enhancing our understanding of human nutrition and body physiology. In the realm of plants this knowledge can mean broadening the gene pool for hybridization, developing healthier and more nutritious plants, creating more beautiful surroundings, or averting worldwide famine. The future of biotechnology is in our hands, and it is our obligation to see that every reasonable precaution is exercised so that the products of this new science only benefit society.

CULTURE METHODS

The following are culture methods that not only have the potential for plant regeneration but also have biotechnological applications in such areas as physiology, nutrition, disease and herbicide resistance, and the production of secondary products. The requirements for these procedures are generally more sophisticated and demanding than for basic tissue culture and may not be realized by even the most astute students or skilled laboratory scientists.

CALLUS CULTURE

Historically, carrot *(Daucus carota)* and tobacco *(Nicotiana)* callus cultures have been the classic models for plant tissue culture. Regeneration of whole plants from callus was standard

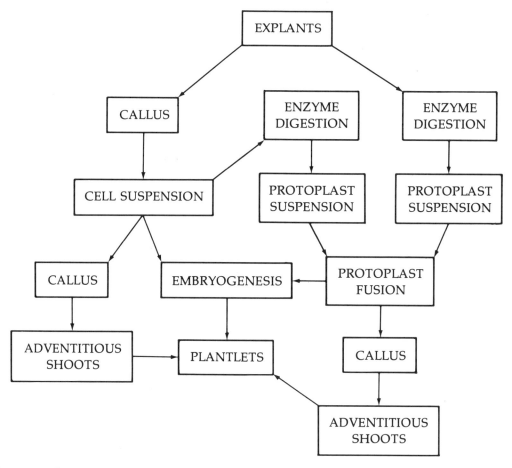

Figure 13-6. Summary cell culture flow chart.

procedure until the merits of shoot tip culture were realized; a callus phase usually was induced before obtaining shoots and roots. For some plants the callus stage is still required, particularly in the grass family (Gramineae). Callus culture seems to produce more mutants than does shoot culture, but sometimes mutants are deliberately induced for genetic variability. Careful management can minimize the potential for mutations in callus, primarily by limiting the length of time as callus and by frequent transfer. Callus is a source of cells for liquid cell suspension and embryo cultures—the ultimate form for mass production, as millions of plants can be started in a very small space. The process of transferring liquid cell and embryo cultures requires only the transfer of an aliquot of liquid, a process which can be automated. To produce carrot *(Daucus carota)* callus is relatively simple.

1. Clean viable seeds in alcohol and bleach.
2. Apply several sterile water rinses.
3. Place on half-strength MS medium in test tubes or petri dishes.
4. After germination, place 5- to 10-in (12.5- to 25-cm) sections of hypocotyl on Gamborg's B5 medium (see Section II, *Catharanthus*) with 0.1 mg/liter of the auxin 2,4-D.
5. Place in the dark at about 77 to 85°F (25 to 29°C).
6. Transfer every 4 weeks until there is sufficient growth for the particular study.

CELL CULTURE

Cell culture technology has many objectives, among them being the study of cell nutrition and metabolism, disease and pesticide resistance, the production of secondary products, somatic hybrids, haploid culture, protoplast isolation and fusion, genetic engineering, mutagenesis, embryogenesis, and other phenomena. Cell suspension cultures are a convenient means of studying cell utilization of and response to different nitrogen and carbon sources. The data gathered from such studies are important not only for cell cultures but also for application to plants in the field or greenhouse. Cell cultures are used for screening cell resistance to introduced pathogens, herbicides, fungicides, salinity, and other factors. The cells that do not succumb to the introduction of such factors can grow on to be whole plants possessing corresponding resistance.

Cell culture is a productive and efficient method of multiplication (barring mutations). The cells and/or embryos are ultimately plated out (spread on agar medium) and grown out, with or without a callus stage. Some cells will give rise to somatic embryos; in other cases, plantlets are obtained only by way of organogenesis, or from callus-derived adventitious shoots.

Cell and cell suspension cultures are derived from callus cultures (see Figure 13-6). Cell suspension cultures are more readily obtained if the callus is friable. Such cultures require a rotator or open-platform orbital shaker. Gamborg's B5 is a satisfactory medium for carrot cell suspension culture; 2,4-D is added and agar is omitted. Antibiotics are often necessary.

1. Cut 4 to 6 callus pieces, each about 0.5 in (1 cm) square.
2. Transfer the callus pieces to 125-ml Erlenmeyer flasks containing modified Gamborg's B5 medium.
3. Stopper the flasks with cotton and cover tops with foil.
4. Place the flasks on a shaker in continuous light.
5. Subculture every week by allowing the cells to settle and decanting part of the medium, which is then replaced with fresh medium. A cell turbid suspension usually develops within 6 weeks.

SOMATIC EMBRYOGENESIS

Somatic embryogenesis is the formation of embryos from somatic (vegetative) cells. This must not be confused with embryo rescue, which is the removal of an embryo from a seed and using that embryo as an explant for micropropagation. Dividing cells undergoing embryogenesis first form a globular stage, then a heart stage, and then a torpedo stage, from which plantlets develop.

Although somatic embryo cultures and cell cultures tend to produce abnormal plantlets, an increasing number of species are being produced successfully by these methods (Ammirato 1987). A promising potential application for somatic embryogenesis is encapsulated artificial seed (Redenbaugh 1990). Embryos can be encapsulated in a gel to make artificial seeds, then field planted by fluid drilling (field planting of seeds in water). Celery (*Apium graveolens*), carrots (*Daucus carota*), and bananas (*Musa*) are among the crops that have been planted using encapsulated embryos, with varied success.

To condition carrot cells in suspension for embryo development, apply the following protocol.

1. Decant the cell suspension medium and replace with fresh Gamborg's B5 liquid medium.
2. Continue agitation on the shaker.
3. In 2 weeks, decant the medium again.

4. Spread the cells on half-strength Gamborg's B5 or MS medium with agar and without hormones in test tubes or petri dishes.

PROTOPLASTS

Protoplasts, which are cells that lack their normal rigid cellulose walls but still retain their plasma membranes, are a valuable research tool in biotechnology and are useful for plant breeders and hybridizers. Because the protoplasts lack walls, many substances are readily taken up through the fragile membranes, including viruses, nuclei, chloroplasts, DNA, proteins, other macromolecules, and other protoplasts. Starting material for protoplast cultures may be any rapidly growing tissue cultured callus or shoots, cell culture, or leaf mesophyll from young, fully expanded leaves. Most growers probably will not want to attempt protoplast culture themselves because it is a very difficult procedure, but it is important to understand the principles in order to appreciate their potential. The following are broad statements of protocol for protoplast isolation assuming *in vitro* starting material.

1. Pretreat cells for 1 to 2 days in the dark.
2. Centrifuge cells in Gamborg's B5 medium.
3. Incubate with filter-sterilized enzyme solution (cellulase and Macerozyme) for 1 to 12 hours.
4. Filter through a 60-micron (µ) screen.
5. Centrifuge.
6. Using a microscope, observe for protoplast release.
7. Add to appropriate culture growing medium.

If the starting material is a leaf, clean in bleach and rinse.

1. Peel off the epidermis.
2. Re-apply the enzyme solution, followed by repeating steps 4–7.

Protoplasts obtained from cell suspension cultures of callus are colored off white. Protoplasts from cell suspension cultures of leaf mesophyll cells are green. When fusing protoplasts from 2 different source plants, it may be useful to prepare the protoplasts by these different methods—from callus and from leaf mesophyll cells—to facilitate observation of the fusion and distinction of the dissimilar cells that are being fused. Protoplasts from 2 different plants can also be distinguished from one another microscopically by staining with different colored non-toxic vital stains.

HAPLOID CULTURE

Haploid cultures are derived from pollen, usually by way of anther culture. Haploid plants are valuable to plant breeders and geneticists because there is only a single set of chromosomes to contend with. If chromosomes of haploid plants spontaneously double or are induced to double using colchicine (an extract from the autumn crocus [*Colchicum autumnale*]), the double haploid plants that grow can be desirable for breeding purposes because the 2 sets of chromosomes are identical. Depending on one's objective, however, the ease and frequency of spontaneous doubling can be considered a drawback with some species. Some species are known to possess a degree of non-haploid pollen that is heterozygous (double, but without identical chromosome sets), and therefore useless. If whole anthers are cultured, somatic cells, which are also heterozygous, are present. Haploid plants are often distinguishable by their smaller size. To determine if cells are haploid, the chromosomes must be counted at an early stage.

Anther culture procedures vary significantly from one species to another, and the failure rate is generally high. The timing of obtaining anthers is critical because the stage of pollen development is crucial to success. For example, a rule of thumb for tobacco plants is that the anthers are ready when the sepals and petals of flower buds are the same length. Buds need to be pretreated overnight at 46°F (8°C), cleaned in alcohol and bleach, and the anthers removed. Placed on an appropriate growth medium under fluorescent light (18-hour photoperiod), plantlets may develop in 6 weeks.

ALGAE CULTURE

Fresh water and marine algae are valued for food, fertilizer, and secondary products. Many algae are microscopic, thus lending themselves to liquid culture. Their interesting life cycles are routinely studied in biology classes. Their beautiful patterns may be an inspiration to artists, but their ecological importance is a nightmare to government officials because they congest waterways, lakes, and streams.

Sterilized, filtered soil water (not all soils work) and Hoagland's solution are common media for fresh water algae. Premixed salts for artificial sea water are available commercially for culture of marine algae.

NEW FORESTS

The demand for timber, pulp, paper, and wood derivatives is increasing as populations grow, but forests the world over are declining. Reforestation has not kept pace with the steadily growing demand for wood products. With the urgent worldwide need for reforestation seedling (preferably from superior trees), tissue culture has done much toward helping the situation. Progress is due in part to the millions of dollars that has been spent on research to solve the many problems of propagating superior forest trees, particularly conifers, via tissue culture.

Embryos, cotyledons, juvenile buds, needles, and fascicles have responded to culturing far more readily than have explants from mature trees. Sommer, Brown, and Kormanik (1975) first reported successful propagation of pine *(Pinus)* plantlets obtained from pine embryos multiplied *in vitro*. Research has been plagued by slow growth, aberrations, poor rooting, vitrification, plagiotropic growth, and genetic instability. Both the difficulty in cleaning older material and the lack of response by mature tissues have repeatedly blocked the way to micropropagation of select conifer trees.

Growers are well acquainted with the problems related to lack of juvenility in rooting cuttings. In many cases the methods applied to retain juvenility or induce rejuvenation for purposes of cutting wood also apply to obtaining juvenile material for explants for tissue culture. Hedging or shearing has long been practiced to retain trees of a particular species in a juvenile state in order to provide cuttings that will root.

Alternatively, rejuvenation of mature stock of some species is sometimes accomplished through sequential cuttings, grafts, or hedging. Another method is to repeatedly apply a cytokinin to appropriate nodal areas. This technique has been known to produce witches'-broom, the shoots of which display some degree of juvenility and consequently respond as tissue culture explants.

One approach employed by timber companies is to multiply full-sib (where both parents are known) seedlings *in vitro* for plantlets to increase the orchard, and then use the seed from that orchard for reforestation. The cost effectiveness of reforesting directly with the tissue cultured plantlets is questionable.

Mature trees are occasionally tissue cultured by the induction of somatic embryogenesis

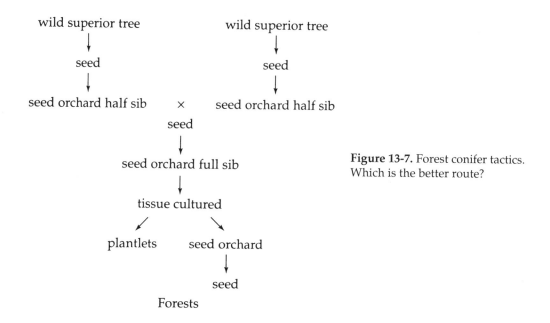

Figure 13-7. Forest conifer tactics. Which is the better route?

from cells of the nucellus. In theory, if the embryo is removed from a seed, the tissues that remain will have the same genetic make-up as that of the female parent plant, not of the embryo. Because of the ease of cleaning seeds and the juvenility of nucellus tissue, this is a viable approach to micropropagation of mature trees. The subsequent cell culture with ensuing somatic embryogenesis is classic procedure.

With respect to hardwood forest trees, the first organogenesis in woody tissue culture was observed by R. J. Gautheret in 1940 when he induced buds from cambial tissues of elm (*Ulmus campestris*). The next hardwood milestone was not reached until 1968, when L. Winton reported the first true plantlets from aspen (*Populus*). To date, other micropropagated forest hardwood trees include *Acacia*, birch (*Betula*), sweet gum (*Liquidambar*), *Paulownia*, *Santalum*, teak (*Tectona*), and *Eucalyptus*. Most of these have been successfully cultured from mature trees, in addition to more routine success from juvenile sources.

Costs of micropropagated transplants have been estimated at several hundred times the cost of conventional production. The gain in quality and production of clonal forests must be weighed against the risks of mutation, disease, or susceptibility to pests.

CRYOPRESERVATION

The world's plant germ plasm historically has been confined to "field genebanks"—the collections of breeders in plantations, orchards, and gardens—and to seed genebanks for storing seeds that lend themselves to drying and maintenance at low temperatures, such as at the U.S. Department of Agriculture's Seed Storage Facility at Fort Collins, Colorado. The concern among conservationists has been that such collections are inadequate.

The work of R. Galzy and of Nag and Street (1973) stimulated an international effort to grow numerous plants *in vitro* and preserve them by *cryopreservation* (storage at about –320°F [–196°C]). The International Board for Plant Genetic Resources (IBPGR) and cooperating laboratories have established a database for research, sources of germ plasm, and other information. They stress the importance of disease indexing and eradication and of the characterization of clones in the various germ plasm repositories. Of special interest is the cryopreservation of sweet potato *(Ipomoea batatas)*, taro or cocoyam *(Colocasia)*, banana or plantain *(Musa)*, coconut *(Cocos)*, sugar cane *(Saccharum officinarum)*, cacao *(Theobroma cacao)*, and *Citrus*.

SECONDARY PRODUCTS

Plants are not as complacent and vulnerable as they may appear. They respond to elicitors (stimulants) by producing secondary metabolites (secondary products). Metabolites are chemicals essential to metabolism, but secondary metabolites are not necessarily required for life. Secondary metabolites often provide the plant's defense and individual response to the environment. Some of these products are *allelopathic*, referring to the phenomenon of chemicals emitted by plants that influence or inhibit the growth of other plants. Perhaps best known is juglone, a toxin produced by black walnut trees *(Juglans nigra)* that is leached from the trees into the ground, where it prevents the growth of some broad-leaved plants.

Many secondary metabolites are of research value or commercial importance, such as pharmaceuticals, medicinals, dyes, food additives, natural flavors, fragrances, gums, and pesticides. The products are conventionally extracted from whole plants or parts of plants. Traditionally the plants are field grown, or they are collected in the wild where they are often in short supply, limited by season and weather, of unreliable yield, and of questionable quality. The collection and removal of plants from the wild is of some concern because of the depletion and potential extinction of certain species and because of the effect of removing native plant species on the natural habitat and environment. The application of plant tissue culture to growing cells, callus, or plantlets for the purpose of extracting secondary products is the object of intensive research. Only as protocols are established and applied to commercial production by farming or cell culture will the wild species enjoy protection.

Tissue culture production provides the opportunity to develop higher yields than is possible from field-grown plants. Higher yields in cell cultures are achieved by adjusting media, selection of high-yield cell lines, genetic engineering, or the application of elicitors, factors that help induce the sought-after chemicals.

The first step toward obtaining secondary products *in vitro* is to grow explants on a solid medium to produce callus. The second phase is to use cells from the callus to proliferate in liquid cell cultures. Although yields from cell cultures can be higher than yields directly from plants, they are, at best, still low. It is necessary to grow large quantities of cells in tanks (reactors or fermentors) in order to produce profitable volumes. The optimal medium, environment, treatment, and reactor must be carefully determined. Extraction involves equipment for chemical separation as well as technical ability, but the premium prices that are paid for secondary metabolites are an attraction; estimates range from hundreds to thousands of dollars per kilogram for certain products, but cost-effective procedures for mass production of these products are costly to develop.

One of the first secondary products to be successfully produced commercially by cell culture was shikonin (shikon), a secondary metabolite from the root of *Lithospermum erythrorhizon* that is used to remedy inflammation, to treat burns and hemorrhoids, as an antibiotic, and as a red dye (Curtin 1983). Mitsui Petrochemical Industries in Japan succeeded in producing this medicinal, which is widely used in Japan but also of interest elsewhere. Because *Lithospermum* cannot be grown commercially in Japan, the company was determined to avoid the need for importing shikonin from China and Korea. By conventional methods *Lithospermum* requires at least 5 years to where the roots will produce a mere 2% shikonin. With careful selection of cell lines, Mitsui was able to obtain 15% of dry weight within a matter of months; their production was estimated at 65 kilograms (143 pounds) per year.

Because of the red color of shikonin, it was relatively easy to select for higher yielding cell lines—the more red color, the greater the product. Mistui determined that modifying the media of Elfried Linsmaier and Folke Skoog and of P. R. White would increase shikonin yield significantly. This increase in shikonin production from *Lithospermum* is achieved by raising the copper levels in the medium (Fujita et al. 1981), by treatment with a fungal homogenate, and by using a special extraction treatment (Chang and Sim 1994). Perhaps the biggest challenge has been to increase the scale of production from agitated liquid media in flasks to larger containers that could still be agitated without injuring the delicate cells.

Three types of bioreactors have been developed that allow for larger scale agitation of media: air-lift fermentors, rotating drums, and stir tanks. An air-lift fermentor provides agitation by introducing filtered air from below, which gently agitates the medium as the bubbles rise to the surface. The stir tanks appear to be the favored method for most large-scale operations (Singh and Curtis 1994). The powerful agitation mechanisms that are used for microbial reactors are too strong for plant cells, so impeller speed must be reduced yet still provide adequate aeration.

E. Ritterhaus and colleagues demonstrated the capability of stir tanks for effective production of secondary metabolites. Using a 75,000-liter tank with an undivided impeller for low shear, they grew *Echinacea purpurea* (coneflower), which has a cell concentration of over 200 grams fresh weight per liter (Singh and Curtis 1994), and thus has a greater output of secondary products. *Echinacea* species contain inulin, betaine, resins, sugars, mineral salts, fatty acids, and echinacein. Echinacin, a commercial extract of *Echinacea*, inhibits the activity of hyaluronidase, a spreading factor enzyme found in snake venom, in some streptococci, in heads of leeches, and in sperm (Windholz et al. 1976).

Oriental ginseng *(Panax ginseng)* and American ginseng *(P. quinquefolius)* are widely used in health foods, cosmetics, pharmaceuticals, and other products, especially in the Orient (Furuya and Ushiyama 1994). Seed germination in these species is poor and plants require 5 years to produce seed. But explants from roots, seeds, or flowers can produce plantlets *in vitro* in 3 months time. With support from the Nitto Deako Corporation in Ibarak, Japan, Tsutomu Furuya and Keiichi Ushiyama were able to produce 500 mg of dried ginseng per day, using 2- and 20-ton bioreactors. Tissue cultured ginseng compared favorably to conventionally produced ginseng with respect to treating gastric ailments, hypoglycemia, and blood flow, when tested in rats. Furuya and Ushiyama state that tissue cultured ginseng inhibits ulcer formation, although ginseng from cultivated plants does not. Some dispute the claims of the medicinal value of ginseng, however. Nevertheless, American ginseng *(Panax quinquefolius)* is becoming exceedingly rare in the wild, having been so intensively collected, and both ginseng species are grown commercially.

Taxol, the most-discussed drug for cancer treatment, traditionally is extracted from the bark of yew *(Taxus)* trees, thus thousands of regal giants from Pacific Northwest old growth forests have been destroyed. Following great political and environmentalist pressure, several alternatives have been developed. Other parts of the trees have been found to produce taxol as well as the bark. Several laboratories are now tissue culturing needles to produce callus for cell culture production of taxol. Meanwhile, other laboratories are striving to synthesize part or all of the complex compound.

When certain cell cultures are treated with elicitors, especially in combination, the yield of secondary metabolites can be increased significantly (Chang and Sim 1994). Among the elicitors that have been discovered to serve this purpose are fungi, ultraviolet radiation, hormone deletion, heavy metal ions, detergents, agar, and heat. Periwinkle *(Catharanthus roseus)* is a plant of particular interest because it is a source of vinblastine and vincristine, alkaloids used in cancer treatment. When a fungal homogenate was added to cell cultures of *Catharanthus roseus* the production of alkaloids was 3 times that of cultures lacking the fungal elicitor.

A filtered extract from the fungus *Pythium* added to a cell culture of *Bidens pilosa* induced the cells to produce secondary products—certain antibiotic phytochemicals—that were not produced when the filtrate was absent. This phenomenon invites speculation not only on the incredible potential of cells, but also on the infinite number of microbial and other inducing agents that exist and the diversity of biochemical products they might produce.

These brief examples of applying tissue culture to obtain secondary products barely touches the surface of this exciting subject. This application of cell culture holds great promise in the areas of nutrition, pesticides, and pharmaceuticals, among others, especially when combined with genetic engineering.

SECTION II

Culture Guide to Selected Plants

Contents of Culture Guide

Ferns —————————————

Oleandraceae
    Nephrolepis, Boston Fern   169
Polypodiaceae, Polypody Family
    Platycerium, Staghorn Fern   169

Conifers —————————————

Cupressaceae, Cypress Family
    Sequoia, Coastal Redwood   170
Pinaceae, Pine Family
    Pinus, Pine   171

Monocotyledons —————————

Amaryllidaceae, Amaryllis Family
    Hippeastrum, Amaryllis   171
    Hypoxis, Star Grass   172
Araceae, Arum Family
    Anthurium, Flamingo Flower   173
    Dieffenbachia, Dumb Cane   173
    Syngonium, Arrowhead Vine   174
Gramineae, Grass Family
    Lolium × *Festuca*, Ryegrass × Fescue   174
Iridaceae, Iris Family
    Freesia   175
    Iris, rhizomatous   175
    Iris, bulbous   176
Liliaceae, Lily Family
    Asparagus   176
    Hemerocallis, Daylily   177
    Hosta, Plantain Lily   178
    Lilium, Lily   178
Orchidaceae, Orchid Family
    Cattleya   179
    Cymbidium   181
    Miltonia, *Odontoglossum*, *Oncidium*   181
    Phalaenopsis   182

Dicotyledons —————————————

Apocynaceae, Dogbane Family
    Catharanthus, Periwinkle   183
Araliaceae, Aralia or Ginseng Family
    Panax, Ginseng   183
Begoniaceae, Begonia Family
    Begonia   184
Berberidaceae, Barberry Family
    Nandina, Heavenly Bamboo   185
Boraginaceae, Borage Family
    Lithospermum   185
Cactaceae, Cactus Family
    Epiphyllum, Orchid Cactus   186
    Mammillaria, Golden Star, Golden Lace
        Cactus   186
Caryophyllaceae, Pink Family
    Dianthus, Carnation   187
Compositae, Composite or Daisy Family
    Chrysanthemum (Tanacetum),
        Pyrethrum   188
    Gerbera, Transvaal Daisy   188
Crassulaceae, Orpine Family
    Kalanchoe, Flaming Katy   189
Cruciferae, Mustard Family
    Brassica, Broccoli   189
Cucurbitaceae, Gourd Family
    Cucumis, Cucumber   190
Ericaceae, Heath Family
    Kalmia, Mountain Laurel   190
    Rhododendron   191
    Vaccinium, Blueberry   192
Gesneriaceae, Gesneria Family
    Saintpaulia, African Violet   192
Juglandaceae, Walnut Family
    Juglans, Walnut   193
Leguminosae, Legume Family
    Leucaena, Lead Tree, White Popinac   195
Myrtaceae, Myrtle Family
    Eucalyptus, Red-Flowering Gum   195
Polemoniaceae, Phlox Family
    Phlox   196
Primulaceae, Primrose Family
    Primula, Primrose   197

saceae, Rose Family
    Actinidia, Kiwi Fruit   197
    Amelanchier, Serviceberry, Saskatoon   198
    Fragaria, Strawberry   198
    Malus, Apple   199
    Prunus, Cherry, Plum   200
    Prunus, Peach   200
    Rosa, Rose   201
    Rubus, Blackberry   202
    Rubus, Raspberry   202
lanaceae, Nightshade Family
    Solanum, Potato   203
taceae, Grape or Vine Family
    Vitis, Grape   203

Introduction to Culture Guide

Generally there are more ways than one to coax an explant or culture to grow. In reading the directions and formulas in this book, the novice might assume that the rules and protocols are binding and infallible; this is certainly not the case. In fact, these instructions are simply what has worked for someone, or close approximations thereof. Occasionally even the very authors of a formula have been unable to repeat their success under different circumstances. When dealing with living organisms, the variables are too numerous to calculate. The guidelines and the basic principles outlined are a place from which to start. The formulas given are derived from the references and other sources of information. You will find more ideas and more references in other texts listed in the bibliography.

Use the procedures described in Section I (especially Chapters 6, 7, and 8) as your guidelines for medium and explant preparation. If you do not have all of the usual additives for a particular formula, leave them out and see what happens. Many plants will grow on $\frac{1}{2}$-strength MS media plus sugar—probably not as well as on a more complete medium but well enough to get by. If a recommended cleaning process calls for 0.1% Tween 20 (which means 1 ml of Tween 20 per liter of water), try a drop or two of household dishwashing detergent instead. If the optimum temperature given is 23°C, most likely the cultures will do just fine in temperatures between 20 and 26°C. You will soon learn which variables work best for you. Tissue culture is not an exact science. Whatever works is the best criteria. Experimentation and flexibility is a vital component of successful tissue culture.

As stated previously in Chapter 8, the 4 stages of culture growth—establishment (Stage I), multiplication (Stage II), rooting (Stage III), and acclimatization or hardening off (Stage IV)—are not necessarily distinct, and the same medium is often used for more than one stage and sometimes for all 3 culture stages. Occasionally no culture medium is required for Stage III because rooting may be done *in vivo*, or perhaps a prerooting medium can be applied in the late-multiplication stage. You may find, again through experimentation and trial and error, that a medium that is specified for one stage may work just as well for the others.

The plants presented are arranged first by the 4 basic categories of ferns, conifers, monocotyledons, and dicotyledons. Within each category, plants are arranged by family and then alphabetically by genus. Scientific and common names are given for all plant families and genera. Within any one family you will often find similar requirements or difficulties in tissue culture.

Murashige and Skoog (MS) formula

Because the formula developed by Murashige and Skoog is used so frequently, it is not itemized in detail for each of the genera in Section II. When the MS medium is required, it is indicated as MS salts, which includes MS major salts, MS minor salts, and iron. The table below gives the standard MS formula; the procedures for making stock solutions are discussed in Chapter 6. If the formula for a particular species calls for 4628 mg/liter of MS salts, you will need 4530 mg of major salts, 33 mg minor salts, and 65 mg iron; if it calls for ½-strength MS, or 2314 mg/liter, the major salts, minor salts, and iron should be halved accordingly.

MS Salts	mg/liter
MAJOR SALTS	
Ammonium nitrate (NH_4NO_3)	1650
Calcium chloride ($CaCl_2 \cdot 2H_2O$)	440
Magnesium sulfate ($MgSO_4 \cdot 7H_2O$)	370
Potassium nitrate (KNO_3)	1900
Potassium phosphate (KH_2PO_4)	170
Subtotal	4530
MINOR SALTS	
Boric acid (H_3BO_3)	6.2
Cobalt chloride ($CoCl_2 \cdot 6H_2O$)	0.025
Cupric sulfate ($CuSO_4 \cdot 5H_2O$)	0.025
Manganese sulfate ($MnSO_4 \cdot H_2O$)	16.9
Potassium iodide (KI)	0.83
Sodium molybdate ($Na_2MoO_4 \cdot 2H_2O$)	0.25
Zinc sulfate ($ZnSO_4 \cdot 7H_2O$)	8.6
Subtotal	32.83
IRON	
Ferrous sulfate ($FeSO_4 \cdot 7H_2O$)	27.8
Na_2EDTA	37.3
Subtotal	65.1
Total mg/liter	4627.93

Refer to Chapter 5 for information on the other chemicals and substances mentioned throughout Section II.

Ferns

Ferns are usually tissue cultured from runners or spores, but spores are seasonal. Spores also take longer because they must go through a gametophyte (sexual) stage.

OLEANDRACEAE, Nephrolepis Family

Nephrolepis exaltata 'Bostoniensis', Boston Fern

Probably more Boston ferns (Nephrolepis exaltata) are propagated by tissue culture than any other ornamental crop. Not only can vast numbers of this houseplant be achieved by tissue culture, but tissue cultured Boston ferns appear healthier, more uniform, and generally more pleasing than those propagated by conventional methods. With a possible production period of less than 4 months, a million plants per year is within reach for even a modest facility.

Explant: 2.5-cm segments of runner tips.

Treatment: Remove 10 cm of actively growing fern runner tips. Cut into 2.5-cm pieces. Stir pieces in 1/10 bleach for 15 minutes. In the transfer hood, dip in 70% alcohol. Rinse in 3 sterile distilled water rinses. Cut into 5- to 10-mm pieces. Place on agar or in liquid medium.

Transfer runner pieces that grow in Stage I to the same medium for Stage II, or they can be chopped up (after removal of the larger fronds) and spread on agar medium of the same composition. The latter procedure is a time-saving process for mass production.

Media: Use ⅔- or ½-strength MS salts with the hormones kinetin and NAA for Stages I and II. Kinetin is omitted in Stage III.

Light: 100 to 300 f.c. from fluorescent light, with photoperiod of 16 hours of light and 8 hours of dark.

Temperature: 25°C (77°F).

References: Cooke 1977a, Cooke 1977b, Cooke 1979, Loescher and Albrecht 1979, Murashige 1974, Murashige and Skoog 1962, Oki 1981, Peterson 1979, Smith 1992.

Nephrolepis media

Compound	Stage I & II	III
	mg/liter	
MS salts	3,086	3,086
Sodium phosphate	125	12.5
Inositol	500	500
Thiamine HCl	5.0	5.0
Kinetin	1.0	–
NAA	0.1	0.1
Sucrose	20,000	20,000
Agar	8,000	8,000
pH 5.7		

POLYPODIACEAE, Polypody Family

Platycerium stemaria, Staghorn Fern

Staghorn ferns (Platycerium stemaria) normally grow on other plants but are not parasitic—they are epiphytes. The roots of staghorn ferns penetrate bark for support, but the plants obtain their nutrients from the air and water. They can be mounted on a piece of bark or wood, on the stems of tree ferns, in moss, or in coarse soil mix or ground bark. Because some of the more desirable species of Platycerium, such as Platycerium wandae, produce very few offshoots (basal shoots) for vegetative propagation, they are good candidates for tissue culture.

Explant: 3-mm shoot apex from offshoot.

Treatment: Obtain offshoots with fronds less than 5 cm wide. Wash in running tap water for 5 minutes. Remove and discard larger fronds and root mass. Excise 1-cm shoot tips. Mix in 1/10 bleach for 10 minutes. Rinse in 3 rinses of sterile distilled water. Remove hairs. Mix in 1/20 bleach for 5 minutes. Excise 3-mm apical dome. Rinse in sterile distilled water. Place on Stage I medium. The explants may turn black overnight, but growth can be expected in 6 weeks. Subdivide and transfer after 2 months. Make further transfers at 3-week intervals.

A time-saving practice for commercial production involves homogenizing Stage III cultures. Place about 40 Stage III plantlets in a sterile blender with 50 ml of sterile distilled water. Blend at low speed for 5 seconds. Aseptically pipet 10-ml portions from the blender onto sterile media in culture dishes, jars, or flasks. In 2 months, transfer the rooted plantlets (possibly as many as 200) to potting mix, tree fern, or sphagnum. Harden off under mist or plastic cover.

Media: Modified MS, with IAA included for Stages I and II but omitted in Stage III. Lily *(Lilium)* medium also gives excellent results for tissue culturing staghorn ferns.

Light: 100 to 300 f.c. from cool-white fluorescent light, 14 hours light/10 hours dark.

Temperature: 25°C (77°F).

References: Cooke 1979, Hennen and Sheehan 1978, Murashige 1974, Murashige and Skoog 1962

Platycerium media

Compound	Stage I & II	III
	mg/liter	
MS salts	4,628	4,628
Sodium phosphate	170	170
Adenine sulfate	80	–
Thiamine HCl	0.4	0.4
Nicotinic acid	1.0	1.0
Pyridoxine HCl	1.0	1.0
IAA	15	–
Sucrose	30,000	30,000
Agar	8,000	8,000
pH 5.7		

Conifers

CUPRESSACEAE, Cypress Family

Sequoia sempervirens, Coastal Redwood

For a conifer, coastal redwood *(Sequoia sempervirens)* lends itself to tissue culture with relatively few problems. Cleaning and growing-on are the most critical steps.

Explant: 1- to 3-cm shoot tips.

Treatment: Wash shoot tips in water with a few drops of liquid detergent. Rinse in sterile distilled water. Rinse in 78% alcohol for one minute. Mix in tap water with 0.1% Tween 20 and 0.1% Captan (fungicide) for 20 minutes. Rinse in sterile distilled water. Mix in 1/10 bleach with 0.1% Tween 20 for 20 minutes. Rinse in 1/100 bleach for 10 seconds. Rinse in 78% alcohol for 10 seconds. Rinse 3 times in sterile distilled water.

Boulay suggests removing 4-cm shoots from the rooting medium, cutting them just above the agar, and soaking in commercial rooting hormone and the fungicide Benlate (125 mg/liter) for 24 hours. He then plants them in perlite/vermiculite (4/1) soil mix.

Media: MS salts with supplements and a trace of IAA suffice for starting and multiplication. Rooting will occur in Stage III when cultures are transferred to ½-strength MS salts, ½ vitamins, lower sugar, no kinetin, both IAA and IBA as auxins, and charcoal. Gelrite is satisfactory for both multiplication and rooting.

Light: 100 to 300 f.c. from cool-white fluorescent light, 16 hours light/8 hours dark. After 4 weeks in rooting, a 2-week period in darkness may help.

Temperature: 25°C (77°F).

References: Ball 1978, Boulay 1979, Murashige 1974, Murashige and Skoog 1962.

Sequoia media

Compound	Stage I & II	III
	mg/liter	
MS salts	4,628	2,314
Sodium phosphate	160	–
Adenine sulfate	80	–
Inositol	100	50
Thiamine HCl	0.4	0.2
IAA	0.5	2.0
IBA	–	3.0

Kinetin	2.0	–
Sucrose	30,000	20,000
Agar (Gelrite)	2,000	2,000
Charcoal	–	600
pH 5.6		

Light: 800 to 1000 f.c. from cool-white fluorescent light, 16 hours light/8 hours dark.

Temperature: 25°C (77°F).

References: Abdullah et al. 1985, Beaty et al. 1985, Sommer and Caldas 1981, Thompson and Zaerr 1981, von Arnold and Eriksson 1981.

PINACEAE, Pine Family

Pinus palustris, Longleaf Pine, Southern Yellow Pine

There is little agreement on media formulas for pines (*Pinus*). Many people use MS in dilute or other modified forms. Rooting and hardening off are especially difficult with pine cultures. As with other genera, requirements may vary with the species, within species, and from clone to clone. Because of these problems, conifers remain one of the most challenging groups of plants to micropropagate.

Explant: Embryos from viable seeds.

Treatment: Agitate seed in 50% bleach for 15 minutes. Rinse 3 times in sterile distilled water. Soak for 40 hours in sterile distilled water. Rinse in 33% bleach for 5 seconds. Rinse in sterile distilled water. Dissect out firm white embryo. Place embryo on Stage I medium.

Media: Low salts and low sucrose prevail in pine media. Some roots will form in Stage II; if not, try rooting in Stage III medium, which is a modified ½-strength MS with NAA and IBA. Alternatively, try adding 10 mg/liter of IBA to Stage II medium for 4 weeks, then return to Stage II without the IBA, to induce root formation

Pinus media

Compound	Stage I	II	III
		mg/liter	
MS major salts	–	–	2,314
Ammonium sulfate	200	–	–
Calcium chloride	150	–	–
Calcium nitrate	–	300	–
Magnesium sulfate	250	740	–
Potassium chloride	300	65	–
Potassium nitrate	1,000	80	–
Sodium phosphate	170	170	–
Sodium sulfate	–	200	–
MS minor salts	32.8	32.8	16.4
MS iron	65.1	65.1	32.6
Inositol	10	10	–
Thiamine HCl	1.0	1.0	–
Nicotinic acid	0.1	0.1	–
Pyridoxine HCl	0.1	0.1	–
IBA	–	–	1.0
NAA	2.0	–	1.0
BA	5.0	–	–
Sucrose	20,000	20,000	10,000
Agar	7,000	7,000	7,000
pH 5.7			

Monocotyledons

AMARYLLIDACEAE, Amaryllis Family

Hippeastrum, Amaryllis

Hippeastrum has been commonly called amaryllis for so long that it is not surprising when it is confused with the genus *Amaryllis*, the belladonna lily, another attractive bulbous plant in the family Amaryllidaceae. *Hippeastrum* is commonly sold as a Christmas gift, potted and attractively packaged, ready for the recipient to water. It is one of a number of bulbous crops that have a relatively low rate of multiplication. Seedlings take 2 to 3 years to flower.

Explant: Shoot tips, ovary tissue, scape slices, young flower buds (grown 10 cm above bulb neck), or 1- to 2-cm twin basal sections of leaves (bulb segments).

Treatment:

Bulbs: Scrub in water with 0.1% Tween 20. Stir in 1/5 bleach for 8 minutes. Agitate in 3 sterile water rinses. Cut 1-cm squares from inner scales. Wash in 1/10 bleach for 2 minutes. Rinse well in sterile water.

Scapes and tight flower buds: Wash in water with 0.1% Tween 20. Rinse by mixing in sterile water for 20 minutes. Dip in 90% alcohol for 10 seconds. Agitate for one minute in sterile water. Mix for 15 minutes in 1/10 bleach or 8% calcium hypochlorite. Trim buds. For the scapes, cut slices 3 mm thick and invert on medium.

Media: Modified MS with additives, depending on the reference used.

Light: 100 to 300 f.c. from fluorescent light. A 16- to 24-hour photoperiod is standard. Six weeks in continuous dark followed by cycle of 12 hours light/12 hours dark has worked well in some instances.

Temperature: 15 to 25°C (59 to 77°F).

References: Christie 1985, Hussey 1975, Mii et al. 1974, Murashige 1974, Murashige and Skoog 1962, Seabrook and Cumming 1977, Tombolato et al. 1994, van Aartrijk and van der Linde 1986, Wilkins 1984.

Hippeastrum media

Compound	Stage I & II		III
	Scales	*Other explants*	
		mg/liter	
MS salts	4,628	–	4,628
Ammonium nitrate	–	1,650	–
Calcium chloride	–	440	–
Magnesium sulfate	–	370	–
Potassium nitrate	–	1,900	–
Potassium phosphate	–	300	–
MS minor salts	–	32.8	–
MS iron	–	65.1	–
Adenine sulfate	–	2.5	–
Inositol	100	100	–
Thiamine HCl	0.1	1.0	–
Nicotinic acid	0.5	1.0	–
Pyridoxine HCl	0.5	1.0	–
NAA	5.0	2.5	–
Kinetin	0.5	0.5	–
Glycine	2.0	–	–
Casein hydrolysate	1,000	500	500
Malt extract	–	500	500
Sucrose	30,000	30,000	30,000
Agar	8,000	6,000	6,000
pH 5.5			

Hypoxis rooperi, Star Grass

Hypoxis rooperi is a species from Africa that is related to our North American native *H. hirsuta*, a delightful rockery species. *Hypoxis rooperi* has been used for centuries in Africa as food and as a medicinal. The medicinal qualities of *Hypoxis* species have been gaining recognition in the United States. The tissue culture techniques well might apply to other corms.

Explant: Upper half of corm, approximately 1.5-cm³ pieces.

Treatment: Remove and discard leaves and roots from corm. Cut upper half of corm horizontally into 3 slices. Remove outer layer (epidermis) of tissue. Wash corm slices in running water for 20 minutes. Stir in 98% ethyl alcohol (or 70% isopropyl alcohol) for 5 minutes. Stir in 2 sterile water rinses. Soak in 3% hydrogen peroxide for 1 minute. Stir in 2 sterile water rinses. Stir in 1/10 bleach for 15 minutes. In the transfer hood, rinse the slices well in sterile water. Further cut the corm, making sure each piece contains some of the cambial layer. You can expect about 35 pieces of 1.5-cm³ size from a mature corm.

Media: Standard MS salts with BA and NAA. Sucrose, inositol, thiamine HCl, and casein hydrolysate are other additives. The same medium without BA is used for rooting.

Light: 100 to 300 f.c. from cool-white fluorescent light, 16 hours light/8 hours dark.

Temperature: 25°C (77°F).

References: Murashige 1974, Murashige and Skoog 1962, Page and van Staden 1984, van Aartrijk and van der Linde 1986.

Hypoxis media

Compound	Stage I & II	III
	mg/liter	
MS salts	4,628	4,628
Inositol	100	100
Thiamine HCl	1.0	1.0
NAA	1.0	1.0
BA	1.0	–

Casein hydrolysate	1,000	1,000
Sucrose	30,000	30,000
Agar	8,000	8,000
pH 5.8		

ARACEAE, Arum Family

Anthurium andraeanum, Flamingo Flower

The *Anthurium andraeanum* cultivars 'Calypso', 'Mauna Kea', and 'Trinidad' are among the varieties of flamingo flowers cultivated by tissue culture methods. The disinfecting process described below has proven to be the most successful method of combating a persistent contamination problem. Because *Anthurium* is a tropical plant, which are particularly subject to contamination, they will benefit from indexing, as discussed in Chapter 11.

Explant: Vegetative buds.

Treatment: Cut nodal sections of approximately $1 \times 1 \times 0.5$ cm size, each containing a bud. Soak for 20 minutes in 1/10 bleach with 0.1% Tween 20. Under a microscope, remove leaf coverings and excise buds no larger than 2 mm at their base. Soak buds in 1/10 bleach for 30 to 45 minutes. Rinse in sterile distilled water for 5 minutes.

Media: For Stage I, add 150 ml of coconut milk per liter of medium to MS salts, vitamins, and 20,000 mg/liter sucrose. Transfer 5- or 10-ml aliquots of the liquid medium into test tubes, cap and sterilize. After cooling, aseptically add one dissected bud to each test tube. Place the inoculated tubes on a rotator or wheel and agitate at approximately 2 to 5 rpm for 6 weeks. For Stage II, transfer 2-node shoot sections to the same medium, but without the coconut milk and with agar and 0.2 mg/liter of BA.

Light: Continuous light at 100 f.c. from fluorescent light.

Temperature: 25 to 28°C (77 to 82°F).

References: Kunisaki 1980, Murashige 1974, Murashige and Skoog 1962, Pierik et al. 1974.

Anthurium media

Compound	Stage I	II	III
		mg/liter	
MS salts	4,628	4,628	4,628
Inositol	100	100	100
Thiamine HCl	0.4	0.4	0.4
Nicotinic acid	0.5	0.5	0.5
Pyridoxine HCl	0.5	0.5	0.5
BA	–	0.2	–
Sucrose	20,000	20,000	20,000
Coconut milk	150 ml	–	–
Agar	–	8,000	8,000
pH 5.5			

Dieffenbachia maculata (D. picta), Dumb Cane

Dieffenbachia is among the most attractive of tropical foliage plants. They do well as houseplants with low light, warm temperature, and a humid atmosphere. The sap of this plant is poisonous; take care that children and pets do not chew it. If the tongue comes in contact with the sap, irritation, swelling and loss of speech can occur—hence the common name dumb cane. *Dieffenbachia maculata* is popular because of its large dark green leaves, which are spotted on both sides with white marbling and blotches.

Because tropical foliage plants are highly susceptible to systemic microbial infections, keep *Dieffenbachia* in a dry, clean, lighted atmosphere for 1 to 3 weeks before removing explant tissue. This practice reduces the number of microbial contaminants on the explants. Once growth is established in tissue culture, indexing for microbial and viral contaminants is highly desirable.

Explant: Lateral buds, or meristems.

Treatment:

Lateral buds: Remove lateral buds from the main stem, keeping approximately 3 mm of surrounding nodal stem tissue around each bud. Mix lateral buds in 1/10 bleach with 0.1% Tween 20 for 10 minutes. Rinse well in 3 rinses of sterile distilled water. Aseptically trim each bud so that only one sheath remains over the tip. Leave approximately 1 mm of surrounding stem tissue. Aseptically place the explants upside down in test tubes containing MS initiation agar medium, with one explant per test tube.

Meristems: Cut apical shoot tip with 3 immature leaves surrounding the apical bud. Disinfect shoot tip cuttings as for lateral buds. Place on sterile paper toweling covering the platform of the dissecting microscope. Aseptically remove all but

1 or 2 of the primordial leaves. Using sterile scalpel and tweezers, excise the meristem and adjoining 1 to 2 mm. Transfer the explant to the agar surface in a test tube containing MS initiation medium.

Media: Stages I and II are grown on an agar medium containing MS salts, sodium phosphate, thiamine HCl, glycine, 2iP, IAA, and sucrose. For Stage III, omit the sodium phosphate, glycine, 2iP, and IAA, and add NAA.

Light: 200 to 300 f.c. from cool-white fluorescent light, 16 hours light/8 hours dark.

Temperature: 25 to 26°C (77 to 79°F).

References: Knauss 1976, Litz and Conover 1977.

Dieffenbachia media

Compound	Stage I & II	III
	mg/liter	
MS salts	4,628	4,628
Sodium phosphate	170	–
Thiamine HCl	0.4	0.4
IAA	0.01	–
NAA	–	0.01
2iP	0.1	–
Glycine	2.0	–
Sucrose	30,000	30,000
Agar	8,000	8,000
pH 5.7 to 5.8		

Syngonium podophyllum, Arrowhead Vine

Syngonium podophyllum is one of the most popular houseplants commercially propagated by tissue culture. Interesting variations of arrowhead vines have the potential for economical multiplication through tissue culture.

Explant: 1- to 5-mm shoot tips.

Treatment: Obtain 1- to 2-cm shoot tips from young plants. Mix in 1/10 bleach with 0.1% Tween 20 for 15 minutes. Rinse well in 3 rinses of sterile distilled water. Excise 1 to 5 mm of shoot apex.

Media: Stages I and II are carried out in stationary or agitated liquid medium with MS salts, sodium phosphate, vitamins, 2iP, and IAA. For Stage III, use MS salts and agar.

Light: For Stages I and II, 300 f.c. from cool-white fluorescent light; for Stage III, 800 f.c. For all stages, 16 hours light/8 hours dark.

Temperature: 25°C (77°F).

References: Makino and Makino 1977, Murashige 1974, Murashige and Skoog 1962.

Syngonium media

Compound	Stage I & II	III
	mg/liter	
MS salts	4,628	4,628
Sodium phosphate	340	–
Inositol	100	–
Thiamine HCl	0.4	–
IAA	1.0	–
2iP	3.0	–
Sucrose	30,000	–
Agar	–	8,000
pH 5.5		

GRAMINEAE, Grass Family

Lolium multiflorum × Festuca arundinacea, Ryegrass × Tall Fescue

The hybrid of *Lolium multiflorum* and *Festuca arundinacea* is infertile and, therefore, completely dependent on vegetative propagation for reproduction. Grasses (the family Gramineae) and legumes do not lend themselves readily to *in vitro* culture; however, special applications require its use. Tissue culture of grasses is especially pertinent to haploid culture, disease resistance, mutation inducement, and genetic engineering.

Explant: Midveins of leaves or lower ends of internodes and peduncles from plants with emerging panicles (inflorescences).

Treatment: Dip stalks in 70% ethyl alcohol for 10 seconds. Mix in 1/10 bleach for 15 minutes. Rinse in 3 rinses of sterile distilled water for one minute each. Remove leaf sheath. Cut peduncle and internode sections into 3-mm segments.

Media: First, produce callus on modified MS with 2,4-D and NAA. Subculture callus every 3 to 4 weeks. After 3 subcultures, wait 8 weeks, then transfer to differentiation medium, which has reduced sugar, agar, and 2,4-D and no NAA.

Light: 100 to 300 f.c. from cool-white fluorescent light, 16 hours light/8 hours dark.

Temperature: 25°C (77°F).

References: Dale et al. 1981, de Fossard 1976a,

George and Sherrington 1984, Kasperbauer et al. 1979, Murashige 1974, Murashige and Skoog 1962, Torello and Symington 1984.

Lolium × Fescue media

Compound	Stage I & II	III
	mg/liter	
MS salts	4,628	4,628
Thiamine HCl	0.1	–
Nicotinic acid	0.5	–
Pyridoxine HCl	0.5	–
2,4-D	2.0	0.25
NAA	0.1	–
Glycine	2.0	2.0
Sucrose	30,000	10,000
Agar	10,000	5,000
pH 5.7		

IRIDACEAE, Iris Family

Freesia

Freesias are cormous plants noted for their fragrance and early blooming. They are versatile outdoors or as potted indoor plants. They are also valuable as cut flowers. Tissue culture of *Freesia* can improve production rates over conventional division by corms.

Explant: 2-mm transverse slices of aerial corms.

Treatment: Agitate aerial corms in 95% ethyl alcohol for 2 minutes. Mix in 1/10 bleach with 0.1% Tween 20 for 10 minutes. Rinse in 3 sterile distilled water rinses for 2 minutes each. Slice 2-mm cross sections. Place on agar medium in continuous dark.

Growth will occur in about 4 weeks in the dark-grown cultures. The subsequent light period needs about 2 weeks, after which the plantlets can be grown-on in moist peat under fluorescent lights and 20°C (68°F). Water with MS nutrient solution for one week, then with tap water. Plant in soil in greenhouse when plants are 8 weeks out of culture.

Media: MS modified with NAA at 0.1 mg/liter and kinetin at 0.05 mg/liter promotes both shoot and root development.

Light: Continuous darkness until formation of shoots and roots, then place in 100 f.c. from flu-

orescent light with 16 hours light/8 hours dark.

Temperature: 25°C (77°F).

References: Anderson and Meagher 1978, Anderson and Mielke 1985, Meyer et al. 1975, Murashige 1974, Murashige and Skoog 1962, Read and Gavinlertvatana 1976.

Freesia medium

Compound	Stage I, II & III
	mg/liter
MS salts	4,628
Inositol	100
NAA	0.1
Kinetin	0.05
Sucrose	30,000
Agar	8,000
pH 5.8	

Iris, rhizomatous

Iris are slow and difficult in culture, but persistence and patience are duly rewarded.

Explant: 2-mm cross sections of peduncle (flower cluster stalks) of young inflorescences.

Treatment: With a sterilized knife, cut young (15-cm) inflorescence. Remove and discard bract tissue. Stir inflorescence in 1/10 bleach with 0.1% Tween 20 for 20 minutes. In the hood, rinse in 3 changes of sterile water for 2 minutes each. Have ready sterile antioxidant solution (0.1% citric acid/0.1% ascorbic acid solution). Discard florets and place peduncles on a sterile paper toweling moistened with antioxidant solution. Slice 2-mm sections of peduncles. Place upside down on agar medium.

Callus will develop from the explant in 6 to 12 weeks. Cut callus into several pieces and place in the light. Transplant the shoots and roots that develop from the edges of the callus pieces (in 2 to 5 months) to peat/perlite (1/1) soil mix. Keep under mist for 2 weeks, then gradually reduce mist and humidity.

Media: Modified MS with added potassium phosphate, casein hydrolysate, malt extract, and adenine sulfate. Add 2.5 mg/liter NAA and 0.5 mg/liter kinetin for callus formation.

Light: Continuous dark for 6 to 12 weeks, followed by 100 to 300 f.c. from fluorescent light with 16 hours light/8 hours dark.

Temperature: 26°C (79°F).

References: Hussey 1975, Hussey 1976, Meyer et al. 1975, Mielke and Anderson 1989, Murashige 1974, Murashige and Skoog 1962.

Iris (rhizomatous) medium

Compound	Stage I & II mg/liter
Ammonium nitrate	1,650
Calcium chloride	440
Magnesium sulfate	370
Potassium nitrate	1,900
Potassium phosphate	300
MS minor salts	32.8
MS iron	65.1
Adenine sulfate	160
Inositol	100
Thiamine HCl	0.4
NAA	2.5
Kinetin	0.5
Casein hydrolysate	500
Malt extract	500
Sucrose	30,000
Agar	6,000
pH 5.5	

Iris, bulbous

The presence of viruses significantly lowers the quality of bulbs. Because virus infection plagues the *Iris* bulb industry, disease-free bulbs bring premium prices. *Iris xiphium* is a bulbous iris that has been successfully tissue cultured.

The first transfer in multiplication occurs at about 6 weeks. Two shorter transfer periods should see the formation of 1- to 2-mm buds. At this stage it is desirable to have a commercial plant pathology laboratory determine if viruses are present. Transfer the individual 1- to 2-mm buds to multiplication medium. A single cycle of 3 to 5 weeks in the intermediate medium is all that is required for conditioning for bulbing. Following 6 months in the bulbing medium, plant in a cool (10°C [50°F]) greenhouse for rooting.

Explant: 0.5- to 1.5-mm shoot tips.

Treatment: In September trim away all but the basal plate with the tiny sheath leaves covering the new shoot tip (0.5 to 1.5 mm). Cut away and discard the basal plate, saving only the shoot tip.

Stir in sterile water with 0.1% Tween 20 for 10 minutes. Rinse in sterile distilled water. Mix in 1/10 bleach for 15 minutes. Rinse thoroughly in sterile water.

Media: Three stages of medium are used: multiplication, intermediate, and bulbing. The first medium is modified MS with BA and IAA; the second omits adenine sulfate, BA, and IAA but adds kinetin and NAA and increases the sucrose; the bulbing stage medium omits the hormones.

Light: For the multiplication and intermediate stage, 100 to 300 f.c. from fluorescent light, 16 hours light/8 hours dark. For the bulbing stage continuous dark for 6 months.

Temperature: Multiplication and intermediate stage, 20°C (68°F); bulbing stage, 25 to 28°C (77 to 82°F).

References: Anderson and Mielke 1985, Meyer et al. 1975, Mielke 1984, Mielke and Anderson 1989, Murashige 1974, Murashige and Skoog 1962.

Iris (bulbous) media

Compound	Multiplication	Intermediate mg/liter	Bulbing
MS salts	4,628	4,628	4,628
Adenine sulfate	80	–	–
Inositol	100	100	100
Thiamine HCl	0.4	0.4	0.4
IAA	1.0	–	–
NAA	–	0.1	–
BA	1.5	–	–
Kinetin	–	0.01	–
Sucrose	30,000	60,000	60,000
Agar	6,000	6,000	6,000
pH 5.7			

LILIACEAE, Lily Family

Asparagus officinalis, Asparagus

Normally, asparagus (*Asparagus officinalis*) from seed is too varied for optimum commercial production, and multiplication by division of field plants is too labor intensive. Asparagus fields ideally consist of only male plants because they produce more and better stalks, and no seeds are prouced that would grow undesirable female volunteers. Some male plants are hermaphroditic and produce "super males," which are ideal subjects for tissue culture.

Tissue culture of asparagus is not easy. Response tends to be low and erratic and very clone dependent. It is largely a 2-step process: first to grow out the buds; then to induce nodal microcuttings to grow crowns with storage and nonstorage roots.

Explant: Lateral buds.

Treatment: Agitate 15- to 20-cm long spears in 1/10 bleach for 10 minutes. Wash in distilled water for 5 minutes. Excise lateral buds and place in medium for fibrous root development. The cladophylls (stems that function as leaves) that grow from the buds may be divided into 3- to 4-cm pieces. Transfer every 3 weeks. Fibrous roots should develop in 4 to 6 weeks. Transfer plantlets that have developed fibrous roots to medium for storage root development. When these have developed a crown and both fibrous and storage roots, which takes 6 to 9 months, transfer to multiplication/division medium. Each transferred segment must consist of a piece of crown with 2 or 3 shoots and both kinds of roots. In this stage, transfer every 6 to 8 weeks; alternate the medium with fibrous root medium

Media: Modified MS, with hormones NAA 0.04 to 0.1 mg/liter), kinetin (0.1 mg/liter), IAA (up to 0.4 mg/liter), and 2iP (0.04 to 0.2 mg/liter). The high levels of inositol, thiamine HCl, nicotinic acid, and pyridoxine HCl are unique. Clonal differences make it difficult to select the optimum amounts of hormones and vitamins. Chin (1982) found that adding ancymidol (A-Rest), at 5 mg/liter, promoted stronger plantlets, reduced height, and helped prevent callus.

Light: 100 f.c. from cool-white fluorescent light, 16 hours light/8 hours dark. For final stage, use 500 to 1000 f.c.

Temperature: 29°C (85°F).

References: Boyd 1995, Chin 1982, Kunachak et al. 1987, Murashige 1974, Murashige et al. 1972, Murashige and Skoog 1962, Yang and Clore 1973, Yang and Clore 1974.

Asparagus media

Compound	Fibrous roots/ crown	Storage roots/ crown	Multiplication/ division
		mg/liter	
MS salts	4,628	4,628	4,628
Adenine sulfate	40	40	40
Inositol	400	400	400
Thiamine HCl	4.0	4.0	4.0
Nicotinic acid	20	20	20
Pyridoxine HCl	20	20	20
IAA	–	0.4	0.4
NAA	0.1	0.04–0.1	0.04–0.1
2iP	–	0.2	0.04–0.2
Kinetin	0.1	–	–
Ancymidol	5.0 ml	5.0 ml	5.0 ml
Sucrose	30,000	30,000	60,000
Agar	5,000	5,000	6,000
Charcoal	–	–	0.0–1.5
pH 5.7			

Hemerocallis, Daylily

Daylily *(Hemerocallis)* meristems cannot be used as explants because they are difficult to clean and the plant dies if they are removed. Transplant plantlets to sand/loam/peat/perlite (2/1/1/1) soil mix. Water with Hoagland's mist for 10 days.

Explant: 2-mm sections of young (10-cm) inflorescence scapes (flower stalk); flower petals or sepals from 1-mm long flower buds.

Treatment: Remove and discard bract tissue from flower buds. Wash inflorescence in water with 0.1% Tween 20. Rinse in sterile distilled water. Mix in 1/10 bleach with Tween 20 for 20 minutes. Rinse 3 times in sterile distilled water. In the transfer hood, moisten a sterile paper towel with sterile antioxidant solution. On the towel, cut the stalk into 2-mm sections. Place sections upside down on agar medium in a test tube.

Media: Modified MS with increased potassium phosphate, plus casein hydrolysate, malt extract, and adenine sulfate. Added hormones are NAA at 10 mg/liter and kinetin at 0.1 mg/liter for callus formation; the NAA is lowered to 0.5 mg/liter for plantlet formation.

Light: Continuous dark for 4 to 8 weeks for callus, followed by 300 to 1000 f.c. continuous light for plantlet development (2 to 6 months).

Temperature: 26°C (79°F).

References: Fitter and Krikorian 1985, Griesbsach 1989, Heuser and Apps 1976, Heuser and Harker 1976, Meyer 1976.

Hemerocallis media

Compound	Callus	Plantlets
	mg/liter	
Ammonium nitrate	1,650	1,650
Calcium chloride	440	440
Magnesium sulfate	370	370
Potassium nitrate	1,900	1,900
Potassium phosphate	300	300
MS minor salts	32.8	32.8
MS iron	65.1	65.1
Adenine sulfate	160	160
Inositol	100	100
Thiamine HCl	0.4	0.4
NAA	10	0.5
Kinetin	0.1	0.1
Casein hydrolysate	500	500
Malt extract	500	500
Sucrose	60,000	30,000
Agar	6,000	6,000
pH 5.5		

Hosta, Plantain Lily

There is room for experimentation with micropropagation of *Hosta* species and cultivars. The fact that some regenerate directly in light conditions and others require darkness for callus production and/or shoot initiation suggests that the relationship of plant readiness to hormone balance needs further study.

Explant: Flower buds 0.5 to 1 cm long, or scape sections 3 mm long.

Treatment: Cut inflorescence, including about 6 cm of scape. Remove buds and cut scape into 3 pieces. Immerse buds and scape pieces in 70% ethyl alcohol for 30 seconds. Rinse in sterile distilled water for 2 minutes. Mix in 1/10 bleach with 0.1% Tween 20 for 10 minutes. Rinse well in 3 rinses of sterile distilled water. Cut scape pieces into 3-mm sections.

Media: Modified MS with vitamins, NAA (0.5 mg/liter), and BA (2 mg/liter). Reduce BA to 0.1 mg/liter for Stages II and III. Meyer added adenine sulfate and casein hydrolysate for bud

culture in standard light conditions.

Light: Some varieties initiate shoots after weeks of darkness (including *Hosta lancifolia*, *H. plantaginea* 'Royal Standard', *H.* 'Honeybells', and *H.* 'Aoki'), followed by 100 to 300 f.c. from fluorescent light with 16 hours light/8 hours dark. Some species initiate shoots without a period of continuous darkness, such as *H. sieboldiana* from flower buds. Still others produce callus in the dark, followed by shoot growth in standard light conditions.

Temperature: 26°C (79°F); 0 to 1°C (32 to 34°F) for cold storage.

References: Meyer 1980, Murashige 1974, Murashige and Skoog 1962, Papachatzi et al. 1980a, Papachatzi et al. 1980b, van Aartrijk and van der Linde 1986, Zillis and Zwagerman 1979, Zillis et al. 1979.

Hosta media

Compound	Stage I	II	III
		mg/liter	
Ammonium nitrate	1,650	1,650	1,650
Calcium chloride	440	440	440
Magnesium sulfate	370	370	370
Potassium nitrate	1,900	1,900	1,900
Potassium phosphate	300	300	300
Sodium phosphate	170	170	170
MS minor salts	32.8	32.8	32.8
MS iron	65.1	65.1	65.1
Adenine sulfate	160	160	–
Inositol	100	100	100
Thiamine HCl	0.4	0.4	0.4
NAA	0.5	0.5	0.5
BA	2.0	0.1	0.1
Glycine	2.0	2.0	–
Casein hydrolysate	500	500	500
Sucrose	30,000	30,000	30,000
Agar	6,000	6,000	6,000
pH 5.5			

Lilium, Lily

Most commercially micropropagated lilies are the Asiatic hybrids and orientals. *Lilium* cultures can be chilled *in vitro* before planting, but most growers chill *in vivo*. For cold storage of bulblets for spring planting, wash the bulblets then dip in fungicide. Pack in barely dry vermiculite in nesh bags in trays and store in darkness at 0 to °C (32 to 34°F).

Anderson used scale base sections as explant material for culturing *Lilium* 'Red Carpet'. Cultures were grown at 100 f.c., with 16 hours light/8 hours dark, and at 21°C (70°F). They were then planted into Jiffy 7's and covered with glass under 400 to 600 f.c. of light and grown to 1-gram size before chilling. Stimart cultured the Easter lilies *Lilium longiflorum* 'Ace' and 'Nellie White' from scale base sections on Linsmaier and Skoog's medium supplemented with 0.03 mg/liter NAA. The cultures were grown in continuous darkness at 25°C (77°F), followed by 3 weeks at 4°C (39°F) before planting into vermiculite.

Explant: 4- to 5-mm cubes from bulb scale bases. Young axillary buds, young (1-cm) flower buds, young shoots, leaf bases, axillary bulbils, and ovary cross sections (1 to 3 mm) of mature unpollinated flower buds have also been successful.

Treatment: Separate inner bulb scales such that they retain a portion of the basal plate. Wash in tap water with a few drops of detergent. Rinse in running tap water for 15 minutes. Rinse briefly in sterile water. Mix in 1/10 bleach for 10 to 20 minutes. Rinse in 3 rinses of sterile distilled water for one minute each. Cut 4- to 5-mm cubes from base of scales, including some basal plate tissue.

Media: Asiatic hybrids and orientals usually respond well to MS media with ½-strength ammonium nitrate and a minimum of amendments, plus 0.03 mg/liter NAA. Formulas that include 2iP and IAA are strong and better designed for Easter lilies. For bud culture, use MS plus organics with 0.1 mg/liter or 1.0 mg/liter each of BA and NAA.

Light: Both intermittent light and continuous darkness are used, depending on the operator and the cultivar. Continuous darkness is preferred, as it leads to the formation of more bulblets and fewer leaves, which is better for transplanting. Bud cultures are better under 100 to 300 f.c. from fluorescent light, 16 hours light/8 hours dark.

Temperature: 21 to 30°C (70 to 86°F), with the warmer end of the spectrum preferable to the cooler. Chilling temperatures are 1 to 9°C (34 to 48°F) for 4 to 6 weeks for Asiatics and 6 to 10 weeks for orientals.

References: Anderson 1977, Anderson and Meagher 1978, Hussey 1975, McRae 1995, McRae and McRae 1979, Murashige 1974, Murashige and Skoog 1962, Stimart and Ascher 1981a, Stimart and Ascher 1981b, Takayama and Misawa 1980, Takayama et al. 1982, van Aartrijk and van der Linde 1986.

Lilium media

Compound	Stage I & II	III (Optional)
	mg/liter	
Ammonium nitrate	825	825
Calcium chloride	440	440
Magnesium sulfate	370	370
Potassium nitrate	1,900	1,900
Potassium phosphate	170	170
MS minor salts	32.8	32.8
MS iron	65.1	65.1
Adenine sulfate	80	–
Inositol	100	100
Thiamine HCl	0.4	0.4
NAA	0.03	0.03
Sucrose	30,000	30,000
Agar	8,000	8,000
pH 5.7		

ORCHIDACEAE, Orchid Family

Cattleya

Conventional orchid propagation is achieved by the division and cultivation of backbulbs (old bulbous roots), which take many years to grow. If an orchid receives an award at an orchid show, its value increases substantially, but vegetative propagation by backbulbs would never be able to meet the demand fast enough—it would take about 10 years to produce only 6 to 12 commercial-size plants. Breeding pure lines by conventional hybridizing is not an alternative because all the cultivated orchids are genetically complex hybrids (highly heterozygous). Meristem culture makes it possible to rapidly multiply an unlimited number of grand prize winners. When orchid seeds germinate they first produce *protocorms*, bulbous tissue with root hairs. Apical meristems in culture follow a similar development.

Cattleya orchids are *sympodial*, meaning that the growth of each shoot originates in the axils of the scale leaves. The growth of each shoot is terminated after one or two seasons.

Cattleyas are more demanding than some other orchids and are comparatively slow growing; it can sometimes take 2 months on a rotator before greening is observed.

Explant: Axillary buds or meristems from 2- to 5-cm shoots. These are taken from actively growing shoots from backbulbs, or preferably from a new "green" bulb to help avoid contamination and provide better growth. However, green bulbs are scarce since a plant usually produces only one new bulb per year. An explant should be taken before the new shoot is 5 cm long because the meristem ceases to grow at an early stage. If the eyes of a backbulb are poorly developed, place the backbulb in a bag of damp peat moss and allow it to swell before excising the bud.

Treatment: Wash the shoot in water with detergent. Dip the shoot in alcohol for 10 seconds. Mix in 1/5 bleach for 15 minutes. Rinse briefly in sterile distilled water and dry on a sterile paper towel. Under a dissecting microscope, carefully remove the overlapping leaves with frequently sterilized tools. Look for a bud about 2 mm in size at the base of a leaf. Sever the bud just below the point of attachment. Use the bud or excise the meristem.

The cut surfaces of many orchid varieties brown quickly as a result of the oxidation of phenolic compounds. These phenolic compounds are toxic to the plantlets and will leach into the medium, so the cultures should be transferred every 2 or 3 days until the browning and bleeding stops. To help prevent browning, dissect in fresh sterile antioxidant solution or coconut milk.

Place the bud or meristem explant in 5 to 10 ml of liquid medium in a test tube that will fit your rotator; the small amount of liquid provides optimum aeration for the explant. Slowly agitate (at 1 to 2 rpm) the liquid cultures on a rotator or wheel. The initial growth that is desired for multiplication is a small mass of protocorms, which are almost microscopic tuberous structures that may be multiplied or induced to form plantlets. They often show greening within one week and proto-

corm masses should be ready for division and transfer in 2 months, or when they have grown to 0.5 cm. As soon as the mass grows to 1 cm it should be divided and put back into liquid or agar medium. When the cultures are grown on agar, frequent division will upset the orientation, or polarity, of the explants and will delay plantlet formation. Once polarity is established, the cultures put out shoots and roots and mature.

Media: Modified Knudson's C medium. Orchids may be started on agar medium, but they are usually cultured initially in an agitated liquid medium; *Cattleya* cultures are best started in a liquid medium. According to Scully, using an agitated liquid medium will accomplish the following:

- Maximizes protocorm tissue by inhibiting expression of polarity.
- Dilutes the growth inhibitor that is produced in response to excision and sterilization.
- Improves aeration.
- Increases the contact of tissue surface with the nutrients.

Supplement media with coconut milk to produce protocorms and avoid callus. Coconut milk is a frequent addition to orchid media. It may be purchased from many commercial suppliers, or you may extract it yourself from a coconut, as described in Chapter 5. Stage III plantlets may not need coconut milk, and agar is used during that stage.

Cattleyas cannot use nitrates as a nitrogen source because they lack the enzyme, nitrate reductase, that reduces nitrates. Using an ammonium ion, as in ammonium sulfate, is not an answer to the problem either because it renders the pH too low. Therefore, 500 mg of urea is substituted for ammonium sulfate in Knudson's formula for these orchids. *Cattleya* also requires potassium, so 500 mg of potassium chloride is added.

Light: For Stage I, continuous light at 100 to 300 f.c. from cool-white fluorescent light. For Stages II and III, 100 to 300 f.c. either continuous or 16 hours light/8 hours dark.

Temperature: 26°C (79°F).

References: Arditti 1977, Hartmann and Kesser 1983, Knudson 1946, Morel 1964, Morel 1965, Morel 1974, Murashige 1974, Murashige and Skoog 1962, Sagawa and Kunisaki 1982, Scully 1967, Vacin and Went 1949, van Overbeek et al. 1941, Wimber 1963.

Cattleya medium

Compound	Stage I, II & III mg/liter
Calcium nitrate	1,000
Magnesium sulfate	250
Potassium chloride	500
Potassium phosphate	250
MS minor salts	32.8
MS iron	65.1
Urea	500
Sucrose	20,000
Coconut milk	100 ml
Agar (Stage III only)	6,000
pH 5.5	

Cymbidium

Cymbidium orchids are very forgiving in that they are relatively easy to tissue culture. They are sympodial. Cymbidiums have excellent regenerative powers and can be cultured initially on a solid medium, from dormant or active axillary buds, even if the shoots are longer than 5 cm.

Explant: 2- to 5-mm buds from axillary shoots. These may be dormant buds from backbulbs.

Treatment: See *Cattleya*.

Media: For Stage I, either agar or liquid modified Knudson's C medium is preferred, but modified Vacin and Went's *Cymbidium* orchid medium is satisfactory; liquid media may produce faster and more abundant growth. For Stages II and III, use either Knudson's C or Vacin and Went's medium. Vacin and Went's medium has also been used successfully for culture of *Dendrobium* and *Phalaenopsis* orchids.

Light: For Stage I, continuous light at 100 to 300 f.c. from cool-white fluorescent light. For Stages II and III, same light intensity in cycle of 16 hours light/8 hours dark.

Temperature: 22°C (72°F).

References: Arditti 1977, Knudson 1946, Morel 1960, Morel 1964, Morel 1974, Sagawa and Kunisaki 1982, Vacin and Went 1949, Wimber 1963.

Cymbidium media

Compound	Stage I, II & III	
	Knudson's C	Vacin and Went's
	mg/liter	
Ammonium sulfate	500	500
Calcium nitrate	1,000	–
Calcium phosphate	–	200
Magnesium sulfate	250	250
Manganese sulfate · 4H$_2$O	–	7.5
Potassium nitrate	–	525
Potassium phosphate	250	250
MS minor salts	32.8	–
MS iron	65.1	–
Ferric tartrate	–	28
Thiamine HCl	–	0.4
Sucrose	20,000	20,000
Coconut milk	100 ml	100 ml
pH 5.8		

Miltonia, Odontoglossum, and Oncidium

General information on tissue culture of the family Orchidaceae is given under the *Cattleya* entry. The genera of orchids discussed here—*Miltonia, Odontoglossum,* and *Oncidium*—are sympodial.

Explant: Axillary buds or meristems from 2- to 5-cm shoots from backbulbs or new green bulbs. An explant should be taken before the new shoot is 5 cm long because the meristem ceases to grow at an early stage. If the eyes of a backbulb are poorly developed, place the backbulb in a bag of damp peat moss and allow it to swell before excising the bud.

Treatment: See *Cattleya*.

Media: Modified Knudson's C medium. These orchid genera may be started on agar medium, but they are usually cultured initially in agitated liquid medium, which is the preferred method (see *Cattleya*). Coconut milk may not be needed for Stage III plantlets.

Light: For Stage I initiation, continuous light at 100 to 300 f.c. from cool-white fluorescent light. For multiplication, shoot-tip formation, and rooting, either continuous or 16 hours light/8 hours dark at 100 to 300 f.c. from cool-white fluorescent light.

Temperature: For initiation, 26°C (79°F). For multiplication, shoot-tip formation, and rooting, 22 to 25°C (72 to 77°F).

References: Arditti 1977, Hartmann and Kester 1983, Knudson 1946, Morel 1964, Morel 1965, Morel 1974, Murashige 1974, Murashige and Skoog 1962, Sagawa and Kunisaki 1982, Scully 1967, Vacin and Went 1949, van Overbeek et al. 1941, Wimber 1963.

Miltonia, Odontoglossum, and *Oncidium* medium

Compound	Stage I, II & III mg/liter
Ammonium sulfate	500
Calcium nitrate	1,000
Magnesium sulfate	250
Potassium phosphate	250
MS minor salts	32.8
MS iron	65.1
Sucrose	20,000
Coconut milk	100 ml
Agar (Stage III only)	6,000
pH 5.5	

Phalaenopsis

Phalaenopsis are monopodial—the main stem growth continues indefinitely. They are usually tissue cultured from nodes in agar medium.

Explant: Nodal sections or meristems.

Treatment:

Nodal sections: With a clean, single-edged razor blade, cut each node from a flower stalk, leaving 5 to 10 mm of stem on either side. Using tweezers or fine forceps, carefully remove the scale that covers the node. Agitate for 5 minutes in distilled water with several drops of detergent. In the transfer hood, soak nodal sections for 5 minutes in 1/1 solution of bleach and distilled water. Rinse well in 3 sterile distilled water rinses. Place each section in a test tube containing 10 ml agar medium, making sure of good contact with the medium.

Meristem: Remove outer leaves from 3-cm shoots. Dip in ethyl alcohol for 2 seconds, then mix in 1/10 bleach for 15 minutes. Rinse in sterile water. Remove remaining leaves from shoots. Excise meristem consisting of apical dome, 2 leaf primordia, and a cube of tissue, all less than 0.5 mm.

Media: Grow and multiply buds or meristems in agar or liquid medium, using either modified Knudson's C or Vacin and Went's formula with coconut milk. The cultures may be started in agitated liquid, but solid agar medium is usually satisfactory and is required for Stage III.

Light: 150 f.c. from cool-white fluorescent light, 16 hours light/8 hours dark.

Temperature: 24 to 26°C (75 to 79°F).

References: Arditti et al. 1977.

Phalaenopsis media

Compound	Stage I & II Knudson's C	Vacin and Went's	III
		mg/liter	
Ammonium sulfate	500	500	500
Calcium nitrate	1,000	–	1,000
Calcium phosphate	–	200	–
Magnesium sulfate	250	250	250
Manganese sulfate · 4H$_2$O	–	7.5	–
Potassium nitrate	–	525	–
Potassium phosphate	250	250	250
MS minor salts	32.8	–	32.8
MS iron	65.1	–	65.1
Ferric tartrate	–	28	–
Thiamine HCl	–	0.4	–
Sucrose	20,000	20,000	20,000
Coconut milk	100 ml	100 ml	–
Agar	–	–	6,000
pH 5.8			

Dicotyledons

APOCYNACEAE, Dogbane Family

Catharanthus roseus, Madagascar Periwinkle

Related to the garden periwinkle *(Vinca)*, Madagascar periwinkle *(Catharanthus roseus)* is a popular subject of research because of its production of secondary metabolites. Among these are ajmalicine (used to help circulatory problems), serpentine and catharanthine (alkaloids used in cancer treatment), and two products used in treatment of leukemia, vinblastine and vincristine. Alkaloid production has been enhanced in *Catharanthus* by osmotic stress, a fungal homogenate *(Aspergillus niger)*, and trans-cinnamic acid (Chang and Sim 1994). Physical and chemical extraction of products from *Catharanthus* cell cultures have been well detailed by Morris et al. (1985).

Explant: Stem and root pieces, or seeds.

Treatment: Scrub under running tap water. Place in alcohol for 10 seconds. Stir in 1/10 bleach for 20 minutes. Rinse in 3 rinses of sterile distilled water. Cut into 1-cm pieces. Rinse again in sterile water. Place on Gamborg's B5 medium with agar and transfer every 14 days. Friable callus should be transferred to agitated liquid medium (agitated at 150 rpm). Transfer callus culture every 14 days by pipetting 20 ml of cells to 100 ml of medium. Filter cells through sterile filter paper and place in serpentine production medium.

Media: The medium for growth is basically Gamborg's B5 plus hormones. The major difference between Gamborg's B5 and MS is that the former uses sodium phosphate instead of potassium phosphate and ammonium sulfate instead of ammonium nitrate. The medium for serpentine production is ½-strength MS with additives.

Light: Continuous dark.

Temperature: 25°C (77°F).

References: Chang and Sim 1994, Curtin 1983, Morris et al. 1985, Preil 1991, Zenk et al. 1977.

Catharanthus media

Compound	Growth	Serpentine
	mg/liter	
Ammonium sulfate	134	–
Boric acid	3.0	–
Calcium chloride	150	–
Cobalt chloride	0.025	–
Cupric sulfate	0.025	–
Magnesium sulfate	250	–
Manganese sulfate	10	–
Potassium iodide	0.75	–
Potassium nitrate	2,500	–
Sodium molybdate	0.25	–
Sodium phosphate	150	–
Zinc sulfate	2.0	–
MS salts	–	2,314
Ferrous sulfate	15	–
Na$_2$EDTA	20	–
Inositol	100	–
Thiamine HCl	10	–
Nicotinic acid	1.0	–
Pyridoxine HCl	1.0	–
2,4-D	1.0	–
IAA	–	1.0
BA	–	0.1
Kinetin	0.1	–
Sucrose	30,000	50,000
Agar	–	8,000
pH 5.5		

ARALIACEAE, Aralia or Ginseng Family

Panax ginseng and **Panax quinquefolius,**
Oriental Ginseng and American Ginseng

Ginsengs are perennial herbs that have palmate leaves, grow to 1 m (3.3 ft) in height, and have tuberous roots, which are the source of a widely used tonic, used especially in the Orient. The roots are also valued as charms and both plants are commonly sold as medicinals. The Oriental ginseng *(P. ginseng)* and the American ginseng *(Panax quinquefolius)* are cultivated in the United States, Canada, Korea, China, and Russia.

The protocol described here is largely based on the work of Andrew Wang, who used 4-year-old roots of American ginseng as source material.

Explant: Pith (soft tissue in the center of a root or stem) from 2-mm root slices.

Treatment: Wash the root in running tap water. Agitate in 70% alcohol for 30 to 60 seconds. Rinse lightly in sterile water and agitate in 1/10 bleach for 20 minutes. (Wang did this under vacuum with occasional agitation.) Rinse in 3 rinses of sterile distilled water. Cut 2-mm slices. Using a cork borer, extract 1-cm wide pieces of pith from the slices. Place on callus induction medium.

After 4 weeks, excise callus from mother tissue and transfer to a solid (or liquid) maintenance medium. Transfer every 3 weeks. When sufficient callus has grown, transfer to embryo medium. The maturation of the somatic embryos, which develop in about 3 weeks, can be assessed by the change in the general appearance of the mass or by microscopic examination. Transfer to plantlet medium.

Media: Callus induction medium is solid MS medium with additives, including 2,4-D. The solid (or liquid) maintenance medium contains dicamba for American ginseng and 2,4-D for Oriental ginseng. For embryo maturation, move to solid medium of MS salts with Gamborg's vitamins and additives. Germination of somatic embryos is done on MS medium with low IBA and NAA.

Light: For callus and embryos, continuous darkness. For somatic embryo maturation and plant regeneration, 300 f.c. from cool-white fluorescent light, 16 hours light/8 hours dark.

Temperature: 27°C (80°F).

References: Choi et al. 1982, Furuya and Ushiyama 1994, Gamborg et al. 1976, Murashige 1974, Murashige and Skoog 1962, Nitsch and Nitsch 1969, Wang 1990.

Panax media

Compound	Callus	Maintenance	Embryos	Plantlets
		mg/liter		
MS salts	4,628	4,628	4,628	4,628
Inositol	–	–	100	100
Thiamine HCl	10	10	10	10
Nicotinic acid	–	–	1.0	1.0
Pyridoxine HCl	–	–	1.0	1.0
2,4-D	2.0	–	1.0	–
2,4-D (*P. ginseng*)	–	2.0	–	–
Dicamba (*P. quinquefolius*)	–	2.0	–	–
IBA	–	–	–	0.5
NAA	–	–	0.4	0.1
Kinetin	1.0	–	–	–
Casein hydrolysate	100	100	–	–
Sucrose	30,000	30,000	30,000	30,000
Agar (Difco-Bacto)	8,000	–	7,000	7,000
Gelrite	–	2,000 (or 0)	–	–

pH 5.7

BEGONIACEAE, Begonia Family

Begonia rex, King Begonia

Begonias are popular for their diversity of color, texture, and application, and are especially useful as houseplants. Wax begonias (*Begonia* Semperflorens-Cultorum hybrids) have small (3- to 3.5-cm) flowers that bloom profusely. Tuberous begonias (*Begonia* Tuberhybrida hybrids) are noted for their huge flowers in rich colors. King begonias are noted for their large colorful leaves.

Explant: Leaf petiole segments 5 mm long.

Treatment: Mix for 10 minutes in 1/5 bleach. Rinse 3 times in sterile water.

Media: MS salts, vitamins, and growth hormones. Reduce sugar to 20,000 mg/liter for rooting. Stage III may be omitted, rooting directly from Stage II to potting mix under high humidity.

Light: For Stages I and II, 100 f.c. from cool-white fluorescent light; for Stage III, 300 f.c. Use 16 hours light/8 hours dark for all stages.

Temperature: 25°C (77°F).

References: Mikkelsen and Sink 1978, Murashige 1974, Murashige and Skoog 1962.

Begonia media

Compound	Stage I & II	III
	mg/liter	
MS salts	4,628	4,628
Inositol	100	100
Thiamine HCl	1.5	1.5

Nicotinic acid	0.5	0.5
Pyridoxine HCl	0.5	0.5
NAA	0.1	0.1
BA	0.4	–
Glycine	2.0	2.0
Sucrose	30,000	20,000
Agar	8,000	8,000
pH 5.5		

BERBERIDACEAE, Barberry Family

Nandina domestica, Heavenly Bamboo

Heavenly bamboo (*Nandina domestica*) are desirable evergreen shrubs that are especially noted for the red coloring of their leaves in the fall in cooler areas. They have male and female flowers on the same plant, but they require another plant for pollination and the subsequent bright red berries. 'Royal Princess' and 'Purpurea' (dwarf) are 2 cultivars that have been successfully tissue cultured. The yellow exudate in culture media is due to the alkaloid berberine, as opposed to many other cultures where the exudates are presumed to be phenolics.

Explant: Lateral or terminal buds.

Treatment: Remove outer leaves from 5-cm shoot tips. Wash shoots in warm water with liquid detergent for 10 minutes. Rinse briefly with water. Agitate in 1/10 bleach with 0.1% Tween 20 for 20 minutes. On sterile paper toweling or filter paper, excise buds under a dissecting microscope.

Media: Gamborg's B5 medium is used with MS iron for Stages I and II. MS salts are used at ⅓ strength for Stage III rooting. MS may be used instead of Gamborg's B5 for all stages, but the latter is preferred. Unique to this formula, charcoal is used to help initiation.

Light: 600 f.c. from Gro-Lux lamps, 16 hours light/8 hours dark. (Gro-Lux lamps are not usually specified for tissue culture because cool-white fluorescent lamps are generally satisfactory. However, because 2 different studies of *Nandina* used Gro-Lux, it is suggested here.)

Temperature: 26°C (79°F).

References: Gould and Murashige 1985, Matsuyama 1978, Murashige 1974, Murashige and Skoog 1962, Smith 1983.

Nandina media

Compound	Stage I	II	III
		mg/liter	
MS salts	–	–	1,543
Ammonium sulfate	134	134	–
Boric acid	3.0	3.0	–
Calcium chloride	150	150	–
Cobalt chloride	0.025	0.025	–
Cupric sulfate	0.025	0.025	–
Magnesium sulfate	250	250	–
Manganese sulfate	10	10	–
Potassium iodide	0.75	0.75	–
Potassium nitrate	2,500	2,500	–
Sodium molybdate	0.25	0.25	–
Sodium phosphate	150	150	–
Zinc sulfate	2.0	2.0	–
MS iron	65.1	65.1	–
Inositol	100	100	100
Thiamine HCl	0.4	0.4	0.4
NAA	0.1	0.1	3.0
BA	1.0	1.0	–
Sucrose	30,000	30,000	30,000
Charcoal	1,500	–	1,500
pH 5.7			

BORAGINACEAE, Borage Family

Lithospermum erythrorhizon

As discussed in Chapter 13, shikonin is a secondary product derived from the root of *Lithospermum erythrorhizon*. It is a traditional medicine in the Orient and it also yields a red pigment that has been used in lipstick, among other products. *Lithospermum erythrorhizon* was one of the first plants to be grown by tissue culture for the production of secondary products. Production of shikonin was enhanced dramatically by the synergistic effect produced when a *Lithospermum* cell culture and a *Penicillium* species were grown together in a liquid culture medium.

Explant: 1-cm pieces of leaf, stem, or root.

Treatment: Scrub under running tap water. Place in alcohol for 10 seconds. Agitate in 1/10 bleach for 20 minutes. Rinse in 3 rinses of sterile distilled water. Cut into 1-cm pieces. Rinse again in sterile distilled water. Place on agar growth medium. Transfer every 14 days. Transfer friable callus to liquid growth medium. Transfer callus culture every 14 days by pipetting 20 ml of cells to 100 ml fresh growth medium. Filter cells and transfer to medium for shikonin production.

Media: Two media are used, one to support initial growth (solid then liquid), the other to promote shikonin production. The iron content for the shikonin production medium is provided by EDTA ferric-sodium salt ($C_{10}H_{12}N_2O_8NaFe$), which is premixed; ferrous sulfate and Na_2EDTA may be substituted. Linsmaier and Skoog's medium is an alternative to White's, a modified form of which is given below for growth.

Light: Continuous dark.

Temperature: 25°C (77°F).

References: Chang and Sim 1994, Curtin 1983, Dixon 1985, Fujita et al. 1981, White 1963.

Lithospermum media

Compound	Growth	Shikonin Production
	mg/liter	
Boric acid	1.5	4.5
Calcium nitrate	300	694
Cupric sulfate	0.01	0.3
Magnesium sulfate	750	750
Manganese sulfate	5.0	–
Molybdenum trioxide	0.001	–
Potassium chloride	65	65
Potassium iodide	0.75	–
Potassium nitrate	80	80
Sodium phosphate	21	19
Sodium sulfate	200	1,480
Zinc sulfate	3.0	3.0
Ferric sulfate	2.5	–
EDTA ferric-sodium salt	–	1.8
Thiamine HCl	0.1	–
Nicotinic acid	0.5	–
Pyridoxine HCl	0.1	–
IAA	1.75	1.75
Kinetin	2.15	–
Glycine	3.0	–
Sucrose	20,000	30,000
Agar	8,000	–
pH 5.6		

CACTACEAE, Cactus Family

Epiphyllum chrysocardium, Heart of Gold, Orchid Cactus

The attractive tropical cactus *Epiphyllum chrysocardium* is known as Heart of Gold because of the yellow filaments at the center of its large white flowers. Tissue culture of Cactaceae can play an important role in the rescue of endangered species.

Explant: Stem cuttings 1.5 to 2 cm long.

Treatment: Dip in 30% alcohol. Rinse in sterile distilled water. Mix in 1/10 bleach for 15 minutes. Rinse 3 times in sterile distilled water.

The sunken shoot primordia should emerge in about 2 weeks after placing on agar medium. Allow shoots to grow 5 or more buds before transferring to rooting medium.

Media: Use ½-strength MS salts with inositol, pyridoxine, thiamine, and sucrose. Recommended hormones for Stages I and II are BA and NAA. For rooting, omit BA and NAA and add IBA.

Light: Sylvania Gro-Lux lamps, 16 hours light/8 hours dark.

Temperature: 26°C (79°F).

References: Lazarte et al. 1982, Mauseth 1979, Murashige 1974, Murashige and Skoog 1962.

Epiphyllum media

Compound	Stage I & II	III
	mg/liter	
MS salts	2,314	2,314
Inositol	500	500
Thiamine HCl	1.0	1.0
Pyridoxine HCl	5.0	5.0
IBA	–	0.01
NAA	0.1	–
BA	1.0	–
Sucrose	20,000	20,000
Agar	8,000	8,000
pH 5.7		

Mammillaria elongata, Golden Star, Golden Lace Cactus

With over 200 species of *Mammillaria,* this genus is a favorite of collectors. These cacti are valued as potted houseplants because of their interesting shapes and easy care. Due to careless and unethical pillaging, however, some cacti are classified as endangered species. Perhaps tissue culture can help save a few.

Explant: Tubercles.

Treatment: Mix excised branches for one hour in a fungicide with 0.5% Tween 20 added. Trim off spines. Mix branches in 1/10 bleach for 30 minutes. Remove the tubercles. Rinse in 2 rinses of sterile distilled water for 2 minutes each.

For root initiation, transfer to the greenhouse in pots containing peat/vermiculite/perlite (2/1/1). Cover with plastic. To harden off, over a 2-week period gradually enlarge a pinhole-sized hole in the plastic cover.

Media: MS salts with organics, which includes high sugar, L-tyrosine, inositol, adenine sulfate, and vitamins. Growth regulators are 2iP, IBA, and IAA.

Light: 100 to 300 f.c. from cool-white fluorescent light, 16 hours light/8 hours dark.

Temperature: 27°C (80°F).

References: Johnson and Emino 1979, Mauseth 1979, Murashige 1974, Murashige and Skoog 1962.

Mammillaria medium

Compound	Stage I & II mg/liter
MS salts	4,628
Sodium phosphate	85
Adenine sulfate	80
Inositol	100
Thiamine HCl	30
Nicotinic acid	10
Pyridoxine HCl	1.0
IAA	0.5
IBA	1.0
2iP	10
L-Tyrosine	100
Sucrose	45,000
Agar	8,000
pH 5.5	

CARYOPHYLLACEAE, Pink Family

Dianthus caryophyllus, Carnation

Carnations *(Dianthus)* are commercially meristemmed to insure virus-free stock.

Explant: 1-mm shoot tip.

Treatment: Remove from stock plants vegetative stems with approximately 12 nodes. Remove all leaves longer than 3 mm. Wash in water with a few drops of detergent. Mix in 1/10 bleach with 0.1% Tween 20 for 5 minutes. Rinse 2 times in sterile distilled water for 3 minutes each. Aseptically meristem 1-mm stem tips with 1 or 2 pairs of leaf primordia and place on agar medium.

Media: Stage I medium deviates from standard MS salts and includes kinetin, NAA, inositol, glycine, and casein hydrolysate, among other additives that may be beneficial but perhaps not essential. Standard MS salts are used for Stages II and III, but with other modifications. In Stage II, omit the casein hydrolysate and reduce the amounts of kinetin and NAA. In Stage III, omit kinetin and NAA from the medium, or root 2-cm shoots directly from Stage II into Jiffy 7's.

Light: Continuous light, 200 f.c. from cool-white fluorescent light.

Temperature: 22°C (72°F).

References: Davis et al. 1977, Earle and Langhans 1975, Levy 1981, Mele et al. 1982, Murashige 1974, Murashige and Skoog 1962.

Dianthus media

Compound	Stage I	II	III
		mg/liter	
Ammonium nitrate	400	–	–
Boric acid	1.6	–	–
Calcium nitrate	144	–	–
Magnesium sulfate	72	–	–
Manganese sulfate	6.5	–	–
Potassium chloride	65	–	–
Potassium iodide	0.75	–	–
Potassium nitrate	80	–	–
Potassium phosphate	12.5	–	–
Zinc sulfate	2.7	–	–
MS salts	–	4,628	4,628
Ferrous sulfate	27.8	–	–
Na_2EDTA	32.8	–	–
trans-Cinnamic acid	1.5	1.5	1.5

D-Pantothenic acid

calcium salt	5.0	5.0	5.0
Inositol	50	100	100
Thiamine HCl	0.1	0.1	0.1
Nicotinic acid	0.5	0.5	0.5
Pyridoxine HCl	0.5	0.5	0.5
NAA	0.2	0.1	–
Kinetin	2.0	0.5	–
Glycine	2.0	2.0	2.0
Casein hydrolysate	3,000	–	–
Sucrose	30,000	30,000	30,000
Agar	6,000	6,000	6,000
pH 5.0			

COMPOSITAE, Composite or Daisy Family

Chrysanthemum cinerariifolium (Tanacetum cinerariifolium), Pyrethrum

Pyrethrums *(Chrysanthemum cinerariifolium)* have been multiplied extensively, especially in South America, for extraction of pyrethrum, an insecticide. Others in the Compositae family have responded to these same tissue culture procedures, including *Chrysanthemum leucanthemum (Leucanthemum vulgare), Chrysanthemum parthenium (Tanacetum parthenium), Chrysanthemum segetum, Anthemis arvensis, Calendula officinalis, Hypochoeris autumnale, Hypochoeris radicata, Leontodon autumnalis,* and *Matricaria maritima (Tripleurospermum maritimum).*

Explant: Young flower heads.

Treatment: Wash young, unopened flower heads in water and a few drops of detergent. Rinse 2 times in sterile distilled water. Mix in 5% calcium hypochlorite solution for 30 minutes. Rinse for 30 minutes in sterile distilled water. Cut the heads in 2 longitudinally. Remove and discard bracts, disc flowers, and ray flowers.

Shoots will develop in 3 weeks—most are very short. Dip 1-cm long shoots in IAA (1% in talcum powder) and root in potting mix. Roots form in about 3 weeks.

Media: Chrysanthemum is not an easy tissue culture subject, and therefore media formulas vary considerably. Deviations from the medium suggested here include kinetin instead of BA, increased sucrose (to 50,000 mg/liter), or add coconut milk.

Light: 300 f.c. from cool-white fluorescent light, 14 hours light/10 hours dark.

Temperature: 18°C (64°F) during light cycle, 14°C (57°F) during dark.

References: Hartmann and Kester 1983, Levy 1981, Roest and Bokelmann 1975, Sonneborn et al. 1982.

Chrysanthemum medium

	Stage I & II
Compound	*mg/liter*
MS salts	4,628
Adenine sulfate	80
Inositol	100
Thiamine HCl	0.4
Nicotinic acid	0.5
Pyridoxine HCl	0.5
IAA	0.1
BA	1.0
Sucrose	30,000
Agar	6,000
pH 5.8	

Gerbera jamesonii, Transvaal Daisy

Although *Gerbera* grows easily from seed, in the past it has been tissue cultured to produce large numbers of plants of specific colors. With the current state of the art such that *Gerbera* come true to color from seed, the justification for tissue culture lies primarily in the multiplication of unique individuals.

Explant: 2 cm of emerging shoot tip. Whole young capitulum (flower heads) may also be used; they are more easily sterilized but may be less responsive.

Treatment: Because contamination is a persistent problem, use greenhouse material that is as clean as possible. Obtain emerging shoot tips with 2 to 3 mm of base from the crown. Wash briefly in running water. Mix in 3 changes of sterile water with 1.0% Tween 20 for 10 minutes each. Rinse in 3 sterile water rinses for 2 minutes each. Mix in 1/10 bleach with 0.1% Tween 20 for 30 minutes. Rinse in 3 mixes, for 15 minutes each, in sterile distilled water. Use whole or excise under microscope emerging buds with 1 mm of base, or use 3 mm of shoot tip. Dip for one minute in 1/100 bleach and place on agar medium.

Media: MS salts with 0.1 mg/liter IAA and

2.0 mg/liter BA for Stages I and II; or use 5 mg/liter kinetin instead of BA. For Stage III, increase IAA to 2 to 10 mg/liter. If there are problems with rooting, reduce MS to ⅓ strength for rooting stage. Two weeks in Stage III medium may be sufficient before transplanting to potting mix.

Light: 100 f.c. from Gro-Lux light in Stages I and II, with 16 hours light/8 hours dark. For Stage III, use 1000 f.c. from cool-white fluorescent light.

Temperature: 25°C (77°F).

References: Murashige 1974, Murashige and Skoog 1962, Murashige et al. 1974, Pierik and Segers 1973, Preil and Schum 1985.

Gerbera media

Compound	Stage I & II	III
	mg/liter	
MS salts	4,628	4,628
Adenine sulfate	80	–
Thiamine HCl	30	30
Nicotinic acid	10	10
Pyridoxine HCl	1.0	1.0
IAA	0.1	2.0
BA	2.0	–
Glycine	2.0	–
Sucrose	30,000	20,000
Agar	8,000	8,000
pH 5.7		

CRASSULACEAE, Orpine Family

Kalanchoe blossfeldiana, Flaming Katy

Kalanchoe is a colorful genus popular in mild climates and a houseplant standby. Requiring only 4 to 6 weeks from culture initiation to potted plantlets, *Kalanchoe blossfeldiana* is ideal for classroom tissue culture demonstration.

Explant: Leaf blade sections (1.5 × 1.5 cm), stem sections (2 cm), or shoot tips (2 mm).

Treatment: Wash 3- to 5-cm cuttings in warm water with detergent. Rinse briefly in distilled water. Mix in 1/10 bleach with 0.1% Tween 20 for 15 minutes. Rinse in 3 changes of sterile distilled water for one minute each. Trim to size.

Media: Modified MS with kinetin and IAA. For Stage III rooting, omit kinetin and IAA, and add IBA.

Temperature: 22 to 26°C (72 to 79°F).

Light: 100 to 300 f.c. from cool-white fluorescent light, 16 hours light/8 hours dark.

References: Griffiths 1994, Murashige 1974, Murashige and Skoog 1962, Smith 1992, Smith and Nightingale 1979.

Kalanchoe media

Compound	Stage I & II	III
	mg/liter	
MS salts	4,628	4,628
Sodium phosphate	120	120
Adenine sulfate	80	–
Inositol	100	100
Thiamine HCl	0.4	0.4
IAA	1.0	–
IBA	–	1.0
Kinetin	1.0	–
Sucrose	30,000	30,000
Agar	6,000	6,000
pH 5.7		

CRUCIFERAE, Mustard Family

Brassica oleracea (Italica group), Broccoli

An important application of broccoli *(Brassica oleracea)* tissue culture is the multiplication of parents of a self-incompatible hybrid that may be particularly desirable. These tissue cultured parents can be planted in the field for open-pollinated seed production.

Explant: 2-mm flower buds, or 1-cm lateral shoot tips.

Treatment: Precondition stock plants by holding in a greenhouse at 16°C (61°F) for 3 weeks. Obtain young flower heads bearing 2-mm buds. Stir heads and lateral shoots in 1/10 bleach for 15 minutes. Excise 2-mm buds. Rinse buds and shoots in 1/100 bleach for 15 minutes.

Transplant rooted plantlets to peat/perlite (1/1) soil in trays under intermittent mist for 3 days, at which time humidity can be reduced. Plants are ready to transplant to pots in 7 weeks.

Media: MS salts with sodium phosphate, inositol, thiamine, and adenine sulfate. For Stages I and II, add 1 mg/liter of IAA and 4 mg/liter of 2iP. Omit 2iP for the rooting stage.

Light: For Stages I and II, 100 f.c. from cool-white fluorescent light, with 16 hours light/8 hours dark. For Stage III, increase intensity to 600 f.c.

Temperature: 19 to 23°C (66 to 73°F).

References: Anderson and Carstens 1977, Anderson and Meagher 1977, Murashige 1974, Murashige and Skoog 1962.

Brassica media

Compound	Stage I & II	III
	mg/liter	
MS salts	4,628	4,628
Sodium phosphate	170	170
Adenine sulfate	80	80
Inositol	100	100
Thiamine HCl	0.4	0.4
IAA	1.0	1.0
2iP	4.0	–
Sucrose	30,000	30,000
Agar	8,000	8,000
pH 5.7		

CUCURBITACEAE, Gourd Family

Cucumis sativus, Gynoecious (female) cucumber

Gynoecious cucumber hybrids produce all, or primarily, female flowers. Tissue culture offers a convenient method for multiplying these plants that is preferable to cuttings.

Explant: Axillary buds 1 to 3 mm long, taken from one-month-old plants.

Treatment: Place leaf axils in 1/10 bleach for 5 minutes. Rinse in 3 changes of sterile distilled water for 2 minutes each. Excise 1- to 3-mm long axillary buds.

Media: MS salts plus vitamins, NAA, and BA.

Light: 400 f.c. from cool-white fluorescent tubes and incandescent bulbs, 16 hours light/8 hours dark.

Temperature: 25°C (77°F).

References: Handley and Chambliss 1979, Murashige 1974, Murashige and Skoog 1962, Wehner and Locy 1981.

Cucumis media

Compound	Stage I & II	III
	mg/liter	
MS salts	4,628	4,628
Inositol	100	100
Thiamine HCl	1.0	–
Nicotinic acid	0.5	0.5
Pyridoxine HCl	1.0	–
NAA	0.1	–
BA	1.0	–
Sucrose	30,000	–
Agar	7,000	–
pH 5.5		

ERICACEAE, Heath Family

Kalmia latifolia, Mountain Laurel

Mountain laurel *(Kalmia latifolia)* is a praiseworthy shrub and a pleasant accompaniment for *Rhododendron,* especially in moderate moist climates. *Kalmia latifolia* is the plant that was used by Lloyd and McCown in developing and publishing their woody plant medium (WPM), which has proved so very useful. The multiplication rate for *Kalmia* using WPM over a 4-week period has been reported at 8 times. Stage III is optional, and plantlets may be rooted directly in 100% peat, at 30°C (86°F), high humidity, and continuous light. *Kalmia* can also be cultured by following the directions and media for *Rhododendron.*

Explant: 2- to 3-cm shoot tips.

Treatment: Carefully remove all leaves more than 1 cm in length. Dip in 70% ethyl alcohol. Mix in 1/10 bleach with 0.1% Tween 20 for 15 minutes. Rinse in 3 rinses of sterile distilled water for one minute each. Trim away any damaged material. Place the explants in agitated liquid WPM for one week; change the medium daily. After one week, transfer to stationary test tubes with the explant bases in the liquid medium. After 2 months, place the axillary shoots on agar medium. If bleeding occurs, transfer to fresh medium immediately.

Media: Liquid woody plant medium (WPM) is a significant change from MS. It has low nitrate, high potassium, and low sugar. The minor elements, vitamins, and hormones are fairly standard, with 1 mg/liter 2iP used for multiplication.

Light: Continuous light at 100 to 300 f.c from cool-white fluorescent light.

Temperature: 28 to 30°C (82 to 86°F).

References: Anderson 1975, Anderson 1978, Lloyd and McCown 1980, Murashige 1974, Murashige and Skoog 1962.

Kalmia media (WPM)

Compound	Stage I & II	III
	mg/liter	
Ammonium nitrate	400	400
Boric acid	6.2	6.2
Calcium chloride	96	96
Calcium nitrate	556	556
Cupric sulfate	0.025	0.025
Magnesium sulfate	370	370
Manganese sulfate	22.3	22.3
Potassium phosphate	170	170
Potassium sulfate	990	990
Sodium molybdate	0.025	0.025
Zinc sulfate	8.6	8.6
MS iron	65.1	65.1
Inositol	100	100
Thiamine HCl	1.0	0.1
Nicotinic acid	0.5	0.5
Pyridoxine HCl	0.5	0.5
2iP	1.0	–
Glycine	2.0	2.0
Sucrose	20,000	20,000
Agar	6,000	6,000
pH 5.2		

Rhododendron

Rhododendrons are cosmopolitan plants that adorn gardens where there is acid soil and moderate temperatures. They come in a wide range of colors and leaf sizes. Mostly evergreen, there are close to 800 species to choose from, with sizes ranging from low shrubs to trees. *Rhododendron* responds well to tissue culture, a good means of securing large numbers of plants. In the field, the tissue cultured plants tend to branch more than normal but sometimes produce unwanted callus.

Explant: 2- to 4-cm shoot tips of actively growing plants, with lateral buds showing.

Treatment: Wash 5-cm shoot tips in water with a few drops of detergent. Remove leaves and terminal bud. Dip in 70% alcohol for 10 seconds. Rinse briefly in distilled water. Mix in 1/10 bleach with 0.1% Tween 20 for 20 minutes. Rinse for one minute in 3 rinses of sterile distilled water. Trim away 1 cm from base. Lay tips firmly on agar slant, making sure of good contact between shoot and agar.

Transfer every 2 weeks until growth appears, then every 4 to 6 weeks. As the lateral buds break, the explant stem should be gradually divided in 2 or 3 transfers (see Figure 8-3). By the time the leaves are completely miniaturized, 3- to 4-fold multiplication can be expected every 6 weeks.

Media: Use WPM (see *Kalmia*) or Anderson's *Rhododendron* formula (below), which is a modified MS with lower nitrate and added iron, sodium phosphate, adenine sulfate, 2iP, and IAA. Rooting can be achieved directly from Stage II into peat/perlite (1/1) soil under mist, or use Stage III for a brief conditioning. For Stage III, use ⅓-strength salts, omit 2iP and IAA, and add IBA and charcoal.

Light: 100 to 300 f.c. from cool-white fluorescent light, 16 hours light/8 hours dark.

Temperature: 25°C (77°F).

References: Anderson 1975, Anderson 1978, Kyte and Briggs 1979, Lloyd and McCown 1980, McCullouch and Briggs 1982, Meyer 1982, Smith 1981, Wong 1981.

Rhododendron media

Compound	Stage I & II	III
	mg/liter	
Ammonium nitrate	400	133
Boric acid	6.2	2.0
Calcium chloride	440	147
Cobalt chloride	0.025	0.008
Cupric sulfate	0.025	0.008
Magnesium sulfate	370	123
Manganese sulfate	16.9	5.6
Potassium nitrate	480	160
Sodium molybdate	0.25	0.08
Sodium phosphate	380	127
Zinc sulfate	8.6	2.9
Ferrous sulfate	55.7	18.6
Na$_2$EDTA	74.5	24.8
Adenine sulfate	80	–

Inositol	100	100
Thiamine HCl	0.4	0.4
IAA	1.0	–
IBA	–	5.0
2iP	5.0–15.0	–
Sucrose	30,000	30,000
Agar	6,000	6,000
Charcoal	–	800
pH 4.5 to 5.0		

Vaccinium corymbosum, Highbush Blueberry

Blueberry *(Vaccinium)* bushes are very much in demand, and there are a number of desirable cultivars specific to certain areas. Success in tissue culture of *Vaccinium corymbosum* has been somewhat erratic. More recent research using leaf tissue has provided encouraging results.

Explant: Single-node cuttings from actively growing shoots.

Treatment: Take soft-wood cuttings and trim off the leaves, making sure not to damage the axillary buds. Wash the cuttings in tap water and a few drops of detergent. Rinse in sterile distilled water with 0.1% Tween 20 for 5 minutes. Agitate in 70% alcohol for 30 seconds. Mix in 1/10 bleach with 0.1% Tween 20 for 30 minutes. Rinse in 3 rinses of sterile distilled water. Dip in sterile antioxidant solution (citric acid/ascorbic acid, 1/1). Rinse for 2 minutes in sterile distilled water. Cut 1.5-cm single-node stem pieces so as to have 5 mm of adjoining stem on either side of the axillary bud. Lay explants on agar and press lightly into the agar so that good contact is made between the total length of the stem and the agar. After about 4 weeks, place the elongated shoots (4 cm) on fresh agar. After about 4 more weeks, remove the leaves and score the leaves lightly on the underside (abaxial) with a scalpel and place on fresh medium with the abaxial side up. New shoots should form on the margins and where the leaves were scored. Transfer every 2 to 4 weeks. In the last transfer prior to rooting (i.e., when sufficient material has multiplied), omit the cytokinin 2iP to permit greater shoot elongation. Remove from the culture shoots that are at least 2 to 3 cm long. Root the shoots on peat/perlite/vermiculite (1/1/1) or milled sphagnum.

Media: Half-strength MS salts with vitamins,

2iP, reduced sugar, and Difco-Bacto agar. Omit 2iP in last transfer before rooting microcuttings directly from Stage II to Stage IV. Anderson's *Rhododendron* formula may be used as an alternative.

Light: 200 to 400 f.c. from warm-white and cool-white fluorescent light, 16 hours light/8 hours dark.

Temperature: 24 to 26°C (75 to 79°F).

References: Anderson 1975, Anderson 1978, Billings et al. 1988, Kyte and Briggs 1979, McCullouch and Briggs 1982, Nickerson 1978, Preece and Sutter 1991, Zimmerman and Broome 1980b.

Vaccinium medium

Compound	Stage I & II mg/liter
MS salts	2,314
Inositol	100
Thiamine HCl	0.4
2iP	4.0
Sucrose	20,000
Agar (Difco-Bacto)	4,000
pH 5.0	

GESNERIACEAE, Gesneria Family

Saintpaulia ionantha, African Violet

It is not surprising that African violets *(Saintpaulia ionantha)* are easily tissue cultured in view of the fact that they usually propagate so readily from leaf cuttings. The decision to culture African violets *in vitro* is a matter of economics that must be considered by the individual grower.

Explant: 1-cm^2 leaf pieces, or 2-mm cross sections of petiole.

Treatment: Wash the whole leaf and petiole in water with 1% detergent. Agitate in 1/10 bleach with 0.1% Tween 20 for 10 minutes. Rinse in 3 rinses of sterile distilled water. Cut leaves into 1×1 cm squares and petioles into 2-mm cross sections.

Media: For leaf sections, use a modified MS with 2 mg/liter of IAA and 0.08 mg/liter of BA. Use MS salts with 0.1 mg/liter of NAA and between 0.01 and 0.1 mg/liter of BA for petiole cross sections.

Light: For Stages I and II, 300 f.c. from cool-white fluorescent light, 16 hours light/8 hours

lark. Light intensity for Stage III rooting should
be reduced to 50 to 100 f.c.

Temperature: 25°C (77°F).

References: Bilkey et al. 1978, Cooke 1977a,
Murashige 1974, Murashige and Skoog 1962.

Saintpaulia media

Compound	Leaf	Petiole
	mg/liter	
MS salts	4,628	4,628
Sodium phosphate	170	170
Inositol	100	–
Thiamine HCl	0.4	–
Nicotinic acid	0.4	–
Pyridoxine HCl	0.4	–
IAA	2.0	–
NAA	–	0.1
BA	0.08	0.01–0.1
Sucrose	30,000	30,000
Agar	8,000	8,000
pH 5.5		

(Stage I, II & III)

JUGLANDACEAE, Walnut Family

Juglans regia, English Walnut, Persian Walnut

Micropropagation of walnuts *(Juglans)* is extremely difficult. Success is slow coming primarily
because of problems stemming from contamination, exudates, and lack of response. Thousands of
dollars have been spent on walnut tissue culture
projects, but with few successes.

Such effort is justified in order to avoid the
time-consuming and troublesome work of grafting, to achieve greater consistency in true-to-type
results, and ultimately, to obtain walnut trees that
grow on their own roots. A few walnut cultivars
are now being commercially micropropagated.

Juglans regia, the English or Persian walnut,
is usually grafted on a rootstock, commonly the
hybrid cultivar 'Paradox' *(J. hindsii × J. regia). Juglans nigra,* the eastern black walnut, is prized for
its wood. It is commonly grown from seed, but it
could be enhanced by tissue culture. Some 'Paradox' have been commercially micropropagated.
Mature 'Chandler' and 2-year-old grafted 'Sun-
land' and 'Vina' cultivars have been micropropa-

gated after spraying the branches weekly with BA
(100 mg/liter) and GA$_3$ (50 mg/liter). The shoots
were trimmed to about 4 buds each time they
grew 20 cm.

Explant: Pretreated 25-cm stems from mature
trees, or 3- to 4-cm nodal segments from seedlings.

Treatment:

Mature plants: Wash stems in running tap
water. Cut off leaves, leaving 2-cm petiole bases
(for orientation). Cut stems, under water, into 5-
cm nodal pieces. Wash for 3 hours under running
tap water. Agitate in 1/5 bleach with 2 drops of
Tween 20 for 15 minutes, more or less depending
on the strength of the shoot. Rinse individual stem
pieces in sterile distilled water.

Seedlings: Wash for 3 hours under running
tap water. Agitate in 1/20 household bleach for 10
to 20 minutes. Rinse in 3 distilled water rinses.

For both kinds of starts, transfer every 2 to 3
days for 3 months. Each week remove and discard
1 to 3 mm of base. After the first 2 weeks, transfer
to larger containers, such as Magenta GA-7's or
pint jars, to accommodate leaf growth. New
shoots may be transferred independently when
they are 3 to 4 cm long. After the 3 months, gradually extend the time between transfers to one
week.

Media: Begin with DKW (Driver/Kuniyuki/
Walnut) basal medium (Driver 1985; McGranahan
et al. 1987). After 2 days, transfer to DKW with BA
and IBA hormones. Transfer well-developed
shoots to prerooting medium.

Light: 300 f.c. from cool-white fluorescent
and Gro-Lux (or equivalent) together for continuous light.

Temperature: 22 to 25°C (72 to 77°F).

References: Driver 1985, McGranahan et al.
1987, McGranahan and Leslie 1988, Rodriguez
1982.

Juglans media (DKW)

Compound	Start	After 2 days	Pre-rooting
	mg/liter		
Ammonium nitrate	1,416	1,416	456.2
Boric acid	4.8	4.8	4.8
Calcium chloride	149	149	149
Calcium nitrate	1,968	1,968	634

Cupric sulfate	0.25	0.25	0.25
Magnesium sulfate	740	740	740
Manganese sulfate	33.5	33.5	33.5
Nickel sulfate	0.005	0.005	0.005
Potassium phosphate	265	265	265
Potassium sulfate	1,559	1,559	1,559
Sodium molybdate	0.39	0.39	0.39
Zinc nitrate	17	17	17
Ferrous sulfate	33.8	33.8	33.8
Na$_2$EDTA	45.4	45.4	45.4
Inositol	100	100	100
Thiamine HCl	2.0	2.0	2.0
Nicotinic acid	1.0	1.0	1.0
IBA	–	0.01	0.15
BA	–	1.0	–
Glycine	2.0	2.0	2.0
Sucrose	30,000	30,000	53,000
Agar (Gelrite)	2,400	2,400	2,400
pH 5.4 to 5.7			

Juglans—Somatic embryogenesis

The development of somatic embryos of *Juglans* in agitated liquid culture eliminates a significant amount of transfer work and still produces tremendous numbers of plants.

Explant: Immature walnut fruits, about 2 months after flowering.

Treatment: Surface sterilize the entire walnut by scrubbing in water with detergent. Rinse several times in sterile distilled water. Agitate nut, first in hydrogen peroxide, then in 1/10 bleach, and then in alcohol, each for 10 minutes, rinsing in sterile distilled water after each agitation. In the hood, aseptically dissect the nut and excise the cotyledon tissue. Excise the cotyledon lobes, about 1-cm pieces, and place on DKW medium with additives. Grow in the dark at room temperature (22°C [72°F]) for 2 to 3 weeks. Transfer to DKW without hormones or glutamine. Somatic embryos will develop in 6 to 24 weeks. During this time, transfer subcultures to the same medium every 2 weeks.

Transfer well-defined embryos with developing radicles to DKW without hormones or glutamine, and divide and transfer any brown tissue masses. Place embryo cultures in the cold, 2 to 6°C (35 to 43°F), for 8 weeks to break any dormancy.

Transfer developed embryos to modified White's medium with charcoal, lower sugar, and agar. After 8 weeks, transfer plantlets that have started leaf and root growth to vials containing peat plugs. Water with White's medium. After weeks, transplant to foam cups containing sterilized potting mix. Cover with plastic bags. Apply systemic fungicide twice a week. When plants are 5 to 8 cm, transplant to 25-cm pots covered with plastic bags. Gradually reduce humidity by poking holes in the plastic bags.

Media: DKW medium with additives for embryo initiation and development, eliminating the hormones during development stage. Use modified White's medium for growth.

Light: Grow in darkness until after cold phase. Then use a combination of cool-white fluorescent and plant-growth (Gro-Lux) fluorescent lamps for 16 hours light/8 hours dark.

Temperature: 22 to 25°C (72 to 77°F); 2 to 6°C (35 to 43°F) during 8-week cold phase.

References: Driver 1985, McGranahan et al. 1987, McGranahan and Leslie 1988, Rodrigue 1982.

Juglans media (somatic embryogenesis)

Compound	Initiation	Development	Growth
	mg/liter		
Ammonium nitrate	456.2	456.2	–
Boric acid	4.8	4.8	–
Calcium chloride	149	149	–
Calcium nitrate	634	634	142
Cupric sulfate	0.25	0.25	–
Magnesium sulfate	740	740	360
Manganese sulfate	33.5	33.5	–
Nickel sulfate	0.005	0.005	–
Potassium chloride	–	–	65
Potassium nitrate	–	–	81
Potassium phosphate	265	265	12
Potassium sulfate	1,559	1,559	–
Sodium molybdate	0.39	0.39	–
Zinc nitrate	17	17	–
Ferric sulfate	–	–	2.5
Ferrous sulfate	33.8	33.8	2.5
Na$_2$EDTA	45.4	45.4	45.4
Inositol	100	100	–
Thiamine HCl	2.0	2.0	0.2
Nicotinic acid	1.0	1.0	1.0

Pyridoxine HCl	–	–	0.2
BA	0.01	–	–
IBA	1.0	–	–
Kinetin	2.0	–	–
L-Glutamine	250	–	–
Glycine	2.0	2.0	4.0
Yeast extract	–	–	100
Sucrose	30,000	30,000	5,000
Agar (Gelrite)	2,400	2,400	–
Charcoal	–	–	5,000
pH 5.4 to 5.7			

LEGUMINOSAE, Legume Family

Leucaena leucocephala, Lead Tree, White Popinac

Legumes have lagged behind many other families in tissue culture production. Surprisingly, they are comparatively slow and difficult. *Leucaena* plants are subtropical trees and shrubs prized as ornamentals and for fuel, forage, and biomass. New improved hybrids should respond to these procedures, and thus hasten their distribution.

Explant: Single nodes from vegetative shoots from one-year-old (or older) trees.

Treatment: Dip 3-cm shoot sections in 90% ethyl alcohol. Stir for 5 minutes in 1/5 bleach. Rinse in sterile water. Stir for 5 minutes in 0.1% mercuric chloride or 1/10 bleach. Rinse 3 times in sterile water.

For growing-on, plant in peat/vermiculite/perlite with slow-release nutrients. Cover with polyethylene plastic for 20 days, but uncover for 1 to 2 hours a day. Do not cover after about 3 weeks, by which time there should be new growth.

Media: MS (or ½-strength MS) with inositol and BA for shoot growth. Use ½-strength MS with kinetin and IBA to initiate rooting. A final transfer onto bridges in White's liquid medium helps to condition roots.

Light: 100 to 300 f.c. from cool-white fluorescent light, 16 hours light/8 hours dark.

Temperature: 28°C (82°F).

References: Goyal et al. 1985, Murashige 1974, Murashige and Skoog 1962.

Leucaena media

Compound	Stage I & II	III	Root conditioning
		mg/liter	
MS salt	4,628 (or 2,314)	2,314	–
Boric acid	–	–	1.5
Calcium nitrate	–	–	300
Magnesium sulfate	–	–	720
Manganese sulfate	–	–	5.3
Potassium chloride	–	–	65
Potassium iodide	–	–	0.75
Potassium nitrate	–	–	80
Sodium phosphate	–	–	16.5
Sodium sulfate	–	–	200
Zinc sulfate	–	–	3.0
Ferric sulfate	–	–	2.5
Inositol	100	100	–
IBA	–	3.0	–
BA	3.0	–	–
Kinetin	–	0.5	–
Sucrose	30,000	15,000	15,000
Agar	8,000	8,000	–
pH 5.8			

MYRTACEAE, Myrtle Family

Eucalyptus ficifolia, Red-Flowering Gum

Tissue culture of *Eucalyptus* has particular significance for forest plantations as well as potential for horticulturally desirable species such as *Eucalyptus caesia* and *E. macrocarpa. Eucalyptus radiata* and *E. polybractea* are useful oil-producing species. Other *Eucalyptus* species have been cultured with varied success on the same or different media, but *E. ficifolia* appears to have the most success. Potential major problems include contamination, particularly with adult material, and bleeding, which usually occurs in fresh material. Bleeding can be minimized by soaking in ascorbic acid and an initial incubation period in the dark (3 to 7 days). Rooting and establishment also may face problems. Success rate of rooting can be very low, especially with certain cultivars.

Explant: Nodes or apical buds from seedlings, adult trees, or coppice (juvenile material growing from the base of adult trees, sometimes in response to injury).

Treatment: Cut off leaves, leaving about 1 cm of petiole. Wash nodes in running tap water for one hour. Wash in distilled water with 0.1% Tween 20 for 5 minutes. Agitate in 5% calcium hypochlorite (freshly prepared and filtered) for 20 minutes. Rinse briefly in 3 sterile distilled water rinses. Soak in 0.2% ascorbic acid solution for 2 hours. Rinse in sterile distilled water. Trim to about 1 cm of main stem at node or apex. Place on agar medium and incubate for 24 hours in the dark, then transfer to normal lighting.

Multiple buds may appear in 2 months from the initial explant. During this period there should be several transfers to fresh medium. Roots may appear in 3 weeks after being placed in rooting medium.

Establishment in sand/loam (1/1) soil takes a great deal of care. Humidity should be lowered very gradually. Cover plants with plastic wrap and place in deep shade for 24 hours. Then place in about ¾ shade. After growth has started, increase lighting very gradually up to full sun—leaves are subject to scorching with an abrupt change to brighter light intensity.

Media: Use ½-strength MS with low sugar. Include BA and NAA hormones for multiplication, and IBA for rooting stage.

Light: Place tubed explants in dark for 24 hours. Move to 100 to 300 f.c. from cool-white fluorescent light in 16 hours light/8 hours dark.

Temperature: 25°C (77°F).

References: Barker et al. 1977, de Fossard 1976a, de Fossard 1976b, de Fossard 1977, Hartney 1982, Hartney and Eabay 1984, Murashige 1974, Murashige and Skoog 1962.

Eucalyptus media

Compound	Stage I & II	III
	mg/liter	
MS salts	2,314	2,314
IBA	–	2.0
NAA	0.16	–
BA	0.2	–
Sucrose	20,000	20,000
Agar	6,000	8,000
pH 5.5		

POLEMONIACEAE, Phlox Family

Phlox subulata, Moss Pink

Moss pink *(Phlox subulata),* a popular garden species, forms a perennial evergreen mat or cushion that provides early color in pinks, purples, or white. Tissue culture of phlox can help eliminate diseases that are transmitted by the usual crown division method of multiplication.

Explant: 0.5- to 1-mm shoot tips excised from actively growing shoots.

Treatment: Remove leaves that are over 1 cm long from 2-cm long shoots. Mix shoots in 1/100 bleach for 25 minutes. Rinse in sterile distilled water by dipping briefly in 10 separate rinses. Excise meristem containing 1 or 2 pairs of leaf primordia.

Media: MS salts with Nitsch's additives (biotin, glycine, pyridoxine HCl, thiamine HCl, nicotinic acid, and folic acid), GA₃, and BA. Rooting medium is the same except GA₃ and BA are omitted and NAA is added.

Light: 100 f.c. from Gro-Lux lamps, 16 hour light/8 hours dark.

Temperature: 22 to 26°C (72 to 79°F).

References: Murashige 1974, Murashige and Skoog 1962, Schnabelrauch and Sink 1979.

Phlox media

Compound	Stage I & II	III
	mg/liter	
MS salts	4,628	4,628
Adenine sulfate	40	–
d-Biotin	–	0.2
Folic acid	0.4	–
Inositol	100	100
Thiamine HCl	0.5	0.5
Nicotinic acid	0.5	0.5
Pyridoxine HCl	0.5	0.5
NAA	–	0.5
BA	5.0	–
GA₃	0.004	–
Glycine	2.0	2.0
Sucrose	30,000	30,000
Agar	6,000	6,000
pH 5.5		

PRIMULACEAE, Primrose Family

Primula acaulis (P. vulgaris), Primrose

Wilbur Anderson, the pioneer of *Rhododendron* tissue culture, developed the formula for primrose, especially for double-flowered *Primula acaulis* to increase clonal lines for F_1 hybrid seed production. Primroses are difficult to clean because the shoot tip area is in such close contact with the soil.

Explant: Shoot buds.

Treatment: Remove small division from parent plant. Wash well in tap water with a few drops of liquid detergent or Tween 20. Cut away and discard outer leaves. Carefully excise shoot bud with a few millimeters of base. Agitate excised shoot bud in 1/10 bleach with 0.1% Tween 20 for 20 minutes. Rinse in 1/100 bleach, followed by 4 or 5 washes in sterile distilled water.

Media: Anderson's inorganics and organics (see *Rhododendron*) specially formulated for double *Primula acaulis*. Half-strength salts are used for Stage III.

Light: 100 to 300 f.c. from cool-white fluorescent light, 16 hours light/8 hours dark.

Temperature: 25°C (77°F).

References: Anderson 1984.

Primula media

Compound	Stage I & II	III
	mg/liter	
Ammonium nitrate	400	200
Boric acid	6.2	3.1
Calcium chloride	440	220
Cobalt chloride	0.025	0.013
Cupric sulfate	0.025	0.013
Magnesium sulfate	370	185
Manganese sulfate	16.9	8.5
Potassium iodide	0.30	0.15
Potassium nitrate	480	240
Sodium molybdate	0.25	0.13
Sodium phosphate	380	190
Zinc sulfate	8.6	4.3
Ferrous sulfate	55.7	27.8
Na_2EDTA	74.5	37.3
Adenine sulfate	80	–
Inositol	100	100
Thiamine HCl	0.4	0.4
IAA	1.5	0.5
BA	2.5	–
Sucrose	30,000	30,000
Agar	6,000	6,000
pH 5.7		

ROSACEAE, Rose Family

Actinidia chinensis (A. deliciosa), Kiwi Fruit

The popularity of kiwi fruit (*Actinidia chinensis*) and the acreage devoted to its growth, particularly in Italy and California, has increased dramatically. Because *Actinidia chinensis* is dioecious (male and female are separate), only vegetative propagation is practical. Cuttings take 2 to 3 years to provide field plants, whereas field plants are produced in one year via tissue culture. Female cultivars such as 'Bruno' or 'Hayward' are commonly grown commercially with male pollinators such as 'Male', 'Matua', or 'Tomuri'.

Monette found that 2 bacterial contaminants that survived explant sterilization treatments did not pose a problem to kiwi proliferation or rooting.

Explant: Meristems 0.2 to 0.5 mm long, or apical shoots 2 cm long.

Treatment: Dormant shoots taken in winter should be stored at 1°C (34°F) for later explanting—spray with fungicide for storage.

Meristems: Wash shoots in water and Tween 20 for 20 minutes. Rinse in sterile distilled water. Dip in 78% alcohol for 10 seconds. Rinse again in sterile water. Mix in 1/10 bleach with 0.1% Tween 20 for 15 minutes. Rinse 3 times in sterile distilled water. Again dip in 78% alcohol for 10 seconds. Rinse 2 times in sterile distilled water. Dissect out meristems, sterilizing the instruments frequently.

Apical shoots: Shoots can also be taken in late spring. Cut 3-cm shoot tips and remove expanded leaves. Stir in 6% bleach with 0.1% Tween 20 for 20 minutes. Rinse 3 times in sterile distilled water. Discard 1 cm from shoot base and place remaining tip on agar medium.

Media: For Stage I, use ¾-strength MS with additives. The sugar and inositol amounts are slightly lower than normally prescribed. An agitated liquid medium for Stage II has provided excellent results in some experiments, and therefore is recommended here. Dispense medium into 125-ml flasks and cap with cotton and aluminum foil. Employing an auxin in Stage III has proven to foster excessive callus. Stage III medium, which is ½-strength MS, may well be omitted, and Stage II shoots can be rooted directly in soil mix. Using a preliminary dip of 2 mg/liter IBA may be advantageous for rooting.

Light: 200 to 300 f.c. from fluorescent light, 16 hours light/8 hours dark.

Temperature: 23°C (73°F).

References: Griffiths 1994, Monette 1986, Murashige 1974, Murashige and Skoog 1962.

Actinidia media

Compound	Stage I	II	III (Optional)
		mg/liter	
MS salts	3,471	4,628	2,314
Sodium phosphate	128	170	–
Adenine sulfate	60	80	–
Inositol	75	100	100
Thiamine HCl	0.3	0.4	0.4
IBA	0.023	0.03	–
BA	2.0	2.0	–
Sucrose	22,500	30,000	20,000
Agar	7,000	–	7,000
pH	5.7	5.0	5.7

Amelanchier alnifolia 'Smoky', Serviceberry, Saskatoon

The hardy ornamental shrub *Amelanchier alnifolia* 'Smoky' is a native of the Pacific Northwest, and it is growing in popularity as new cultivars appear in the trade. Though it is among the easier woody plants to tissue culture, it tends to suffer from vitrification and problems of acclimatization.

Harris offers a formula different from the one outlined here. He used higher BA initially (3 mg/liter), followed by an agitated liquid phase, without hormones, to condition shoots for rooting. Shoots were planted in perlite/vermiculite (3/1) soil mix and dampened with household fertilizer and 1 mg/liter IBA. Harris covered the plants with plastic and placed them under cool-white fluorescent light in continuous lighting.

Explant: Apical 2 cm of young, actively growing shoots.

Treatment: Remove all outer leaves. Dip in 78% alcohol. Stir in 7% bleach for 10 minutes. Dip in 78% alcohol. Rinse 3 times in sterile distilled water. Trim to 5 mm tip.

Media: Several simple modifications of MS have been used successfully for *Amelanchier* culture. The one presented here has low BA content to help avoid vitrification. Three-quarter–strength MS is used for Stage I.

Light: 100 to 300 f.c. from cool-white fluorescent light for all stages. For multiplication, 16 hours light/8 hours dark; for rooting and early acclimatization, continuous light.

Temperature: 22°C (72°F).

References: Harris 1984, Harris 1985, Murashige 1974, Murashige and Skoog 1962.

Amelanchier media

Compound	Stage I	II	III
		mg/liter	
MS salts	3,471	4,628	4,628
Sodium phosphate	–	100	–
Inositol	100	100	100
Thiamine HCl	0.4	0.4	0.4
IBA	–	–	1.0
BA	1.0	1.0	–
GA₃	0.05	0.05	–
Sucrose	30,000	30,000	30,000
Gelrite	3,000	3,000	3,000
Charcoal	–	–	0.6
pH 5.7			

Fragaria ×*ananassa (F. chiloensis* × *F. virginiana)*, Garden Strawberry

More commercial strawberries *(Fragaria)* are propagated by micropropagation than any other food crop. In Europe, two leaders in the field of tissue culture, P. Boxus in Belgium and C. Damiano in Italy, demonstrated the feasibility of mass producing virus-free strawberry plants by tissue culture, especially for field mother plants used by strawberry plant growers. Tissue cultured plants tend to produce more runners, which is an advantage for multiplication in the field but may mean less fruit for the first year. *Fragaria* is easy to micropropagate from runner tips or meristems. Sorvari regenerated plantlets from leaf disks.

Plantlets can be stored *in vitro* in the dark at

°C (39°F) for several months before planting in the field. With special pretreatment in dimethyl sulfoxide, meristems can be stored in liquid nitrogen at cryogenic temperatures (–96°C [–141°F]).

Explant: Meristems from young (5-cm) runner tips that have not yet leafed out. If source plants are not certified as virus free, then the 0.5-mm meristem should be excised from apical and lateral buds as described in Chapter 8.

Treatment: Place about 12 runner tips in a beaker with 100 ml distilled water and 2 drops of Tween 20. Cover and stir on stirrer for 5 minutes (or shake by hand in a tightly covered jar). Remove the tips and place them in a clean beaker with 1/10 bleach. Stir for 10 minutes. Stir in 2 rinses of sterile distilled water for 10 seconds each. In the hood, rinse once more in distilled water and place on sterile toweling in a petri dish under a dissecting microscope. Excise meristems as described in Chapter 8.

Media: MS salts and standard vitamins, with BA for Stages I and II. Rooting can be directly from Stage II into a potting mix or in Stage III medium with IBA and charcoal.

Light: 300 f.c. from cool-white fluorescent light, 16 hours light/8 hours dark.

Temperature: 26°C (79°F).

References: Anderson and Haner 1978, Boxus et al. 1977, Damiano 1980, Kyte and Kyte 1981, Murashige 1974, Murashige and Skoog 1962.

Fragaria media

Compound	Stage I & II	III
	mg/liter	
MS salts	4,628	4,628
Inositol	100	100
Thiamine HCl	0.4	0.4
Nicotinic acid	0.5	0.5
Pyridoxine HCl	0.5	0.5
IBA	–	1.0
BA	0.4	–
Sucrose	30,000	30,000
Agar	6,000	6,000
Charcoal	–	800
pH 5.7		

Malus sylvestris, Apple

Apples (*Malus*) are quite particular in their tissue culture requirements, which can vary in small detail from species to species, cultivar to cultivar, and even clone to clone. Some experimentation is usually necessary to determine the precise factors for optimum response. Important variables include concentration of salts (full-, ½-, or ¼-strength) and type and concentration of auxin, whether IBA, IAA, or NAA.

Explant: One- or two-node stem pieces or apical buds from actively growing shoots with lateral buds showing; or use semi-dormant buds.

Treatment:

Nodal sections: Collect 7- to 10-cm shoots from the first growth in spring. Remove and discard leaves, leaving 1 cm of petiole attached to the stem. Wash with vigorous agitation in tap water and 1% Tween 20 for 10 minutes. Rinse in running tap water for 2 minutes. Stir in 1/10 to 1/5 bleach with 0.1% Tween 20 for 10 to 20 minutes—the harder the shoot, the stronger the bleach and/or the longer the time in bleach solution. In the transfer hood, rinse 3 times for a few seconds in sterile distilled water. Cut 1- or 2-node pieces (about 1 cm long), with a short piece of stem above and below the node. Be sure to sterilize your instruments frequently between cuts. Place lower stem stub into agar so that explant is upright and the bud is above the agar level. Explant buds should break and grow in 3 weeks; promptly transfer to Stage II medium. Transfer cultures at least every 3 weeks.

Semi-dormant buds: Collect twigs bearing swollen, greener, axillary buds. Remove a bud similar to a chip bud, i.e., one that has a sliver of stem on top and a basal section of tissue. Wash vigorously in tap water with 1% Tween 20 for 10 minutes. Stir in 1/10 to 1/5 bleach with 0.1% Tween 20 for 10 to 20 minutes, depending on the hardness of the shoot, as above. In the hood, rinse briefly in 3 rinses of sterile distilled water. With frequent sterilization of instruments, remove outermost bud scales and any burned tissue. Place upright in agar medium.

The remaining instructions apply for both types of explant. Transfer to Stage II medium as soon as explant starts growing. Transfer at least every 3 weeks, more often if there is bleeding or no response. In Stage II, in preparing for rooting, etiolate the cultures by placing them in continuous dark (in cardboard boxes) in the growing room. After 2 to 3 weeks, place the cultures in the light for 1 to 2 weeks to green up before rooting. Once they are rooted in Stage III, place the cultures in 3°C (37°F) for 3 to 6 weeks under 100 to 300 f.c. of light before hardening off.

Media: For Stage I, use ½-strength MS salts and full-strength iron, plus inositol and thiamine HCl, with no hormones. Use Difco-Bacto agar because Gelrite may cause vitrification. Stage II usually requires full-strength MS, but BA and IBA requirements vary. Try BA at 1.0 mg/liter and IBA at 0.1 mg/liter, and change to 0.25 mg/liter BA and 0.01 mg/liter IBA if material starts to become succulent or remains short. In Stage III, return to ½-strength MS salts but retain full-strength iron, thiamine HCl, and inositol. If Stage III is omitted, then chill as above before placing in soil.

Light: 100 to 300 f.c. from cool-white fluorescent light, 16 hours light/8 hours dark.

Temperature: 24 to 26°C (75 to 79°F).

References: Dunstan 1981, Lane 1982, Murashige 1974, Murashige and Skoog 1962, Suttle 1983, Suttle 1986, Zimmerman and Broome 1980a.

Malus media

Compound	Stage I	II	III
		mg/liter	
MS major salts	2,265	4,530	2,265
MS minor salts	16.4	32.8	16.4
MS iron	65.1	65.1	65.1
Inositol	208	208	208
Thiamine HCl	2.5	2.5	2.5
IBA	–	0.1	–
BA	–	1.0	–
Sucrose	30,000	30,000	30,000
Agar (Difco-Bacto)	6,000	7,000	6,000
pH 5.7			

Prunus, Cherry, Plum

Explant: One- or two-node stem pieces from actively growing shoots with lateral buds showing.

Treatment: Collect 7- to 10-cm shoots from the first growth in spring. Remove and discard leaves, leaving 1 cm of petiole still attached to the stem. Wash with vigorous agitation in tap water and 1% Tween 20 for 10 minutes. Rinse in running

tap water for 2 minutes. Stir in 1/10 to 1/5 bleach with 0.1% Tween 20 for 10 to 20 minutes—the firmer the shoot, the stronger the bleach and/or the longer the time in bleach solution. In the hood rinse 3 times for a few seconds in sterile distilled water. Cut 1- or 2-node pieces, with a short piece of stem above and below the node. Be sure to sterilize your instruments frequently between cuts. Place lower stem stub into agar so that explant is upright and the bud is above the agar level. Explant buds should break and grow in 3 weeks, promptly transfer to Stage II medium. Cultures should be transferred at least every 3 weeks.

Media: Half-strength MS salts is usually used throughout, with no hormones in Stage I, although BA at 0.5 to 1.1 mg/liter will help lateral bud break on slow tips. For Stage II, add BA at 0.25 to 0.5 mg/liter and IBA at 0.01 mg/liter. Rooting can be directly into soil from Stage II, or use Stage III medium, omitting BA and increasing IBA to 0.2 mg/liter. Yeast extract at 1 mg/liter is sometimes helpful for cherries in Stage II.

Light: 100 to 300 f.c. from fluorescent light, 16 hours light/8 hours dark.

Temperature: 25 to 27°C (77 to 80°F).

References: Dunstan 1981, Lane 1982, Murashige 1974, Murashige and Skoog 1962, Suttle 1983.

Prunus (cherry, plum) media

Compound	Stage I	II	III
		mg/liter	
MS salts	2,314	2,314	2,314
Inositol	208	208	208
Thiamine HCl	2.5	2.5	2.5
IBA	–	0.01	0.2
BA	–	0.25–0.5	–
Yeast extract	–	1.0	–
Sucrose	30,000	30,000	30,000
Agar (Difco-Bacto)	6,000	6,000	6,000
pH 5.7			

Prunus persica, Peach

Considerable work has been done on peaches (*Prunus persica*) with minor deviations from conventional media, but with only modest success. Almehdi and Parfitt have digressed from the norm with seemingly notable success in Stages

I and II using both juvenile and mature explant material. They supported shoot growth and multiplication of 56 varieties using their AP medium. Establishment in soil continues to be a problem, however.

Explant: Actively growing shoot tips, 1.5 cm long.

Treatment: Obtain 1.5-cm long actively growing shoot tips. Immerse in 15% bleach for 30 minutes. Rinse 4 times in sterile distilled water. Trim to 0.5 cm before placing on Gelrite agar.

Media: Based on Gamborg's B5 medium, Almehdi and Parfitt's (AP) medium has less magnesium sulfate, twice the ammonium sulfate and manganese sulfate, and lower vitamin content. BA is added at 6 mg/liter and IBA at 0.01 mg/liter. Prerooting treatment consists of 2 weeks in a liquid medium containing 2 times the multiplication formula. For Stage III, use ½-strength potassium nitrate and ½-strength ammonium sulfate plus up to 1.0 mg/liter of IBA.

Light: For Stages I and II, 200 f.c. from cool-white fluorescent and Gro-Lux lights for 15 hours light/9 hours dark. Reduce light intensity to 140 f.c. for rooting in Stage III.

Temperature: 26°C (79°F).

References: Almehdi and Parfitt 1986, Feliciano and de Assis 1983, Gamborg 1986, Hammerschlag 1982, Miller et al. 1982, Parfitt and Almehdi 1986, Suttle 1983, Zimmerman 1978, Zimmerman 1986.

Prunus persica media

Compound	Stage I & II	Pre-rooting mg/liter	III
Ammonium sulfate	270	540	135
Boric acid	4.5	9.0	4.5
Calcium chloride	150	300	150
Cobalt chloride	0.03	0.06	0.03
Cupric sulfate	0.05	0.1	0.05
Magnesium sulfate	190	380	190
Manganese sulfate	20	40	20
Potassium iodide	0.75	1.5	0.75
Potassium nitrate	2,500	5,000	1,250
Sodium molybdate	0.06	0.1	0.06
Sodium phosphate	150	300	150
Zinc sulfate	2.0	4.0	2.0
MS iron	65.1	130	65.1
Inositol	25	50	25
Thiamine HCl	5.0	10	5.0
Nicotinic acid	0.5	1.0	0.5
Pyridoxine HCl	1.0	2.0	1.0
IBA	0.01	0.02	1.0
BA	6.0	12	–
Sucrose	20,000	30,000	20,000
Agar (Gelrite)	2,000	–	2,000
pH 5.7			

Rosa, Rose

Roses *(Rosa)* are garden favorites in most temperate climates, and new hybrids are continually being offered on the market. Tissue culture is a viable alternative to grafting roses. Unfortunately, many roses are somewhat uncooperative when subjected to tissue culture. Success is dependent on careful technique and the particular clone.

Explant: 5-cm actively growing shoot tips from disease-free plants.

Treatment: Remove all leaves except tiny terminal ones. Wash shoots in water with a few drops of detergent. Mix in 1/10 bleach for 20 minutes. Rinse in 3 changes of sterile distilled water for 2 minutes each. Trim base to make 1- to 3-cm explants. Place shorter explants on modified WPM agar medium and longer explants in modified WPM liquid medium with the lower half of the stem in the liquid. As the terminal and lateral shoots grow out, transfer 2- to 3-cm pieces from the liquid medium to agar medium.

It is often best to omit Stage III, rooting shoots directly from Stage II into sand/peat moss (1/1) potting mix. Place rooted plantlets in potting mix under high humidity. Root initials may grow best if plantlets *in vivo* are grown in the dark for a few days.

Media: Modified WPM agar medium and modified WPM liquid medium. Add BA at 1 mg/liter for Stage II multiplication.

Light: 100 to 300 f.c. from cool-white fluorescent light, 16 hours light/8 hours dark.

Temperature: 25°C (77°F).

References: Davies 1980, Hasagawa 1979, Hasagawa 1980, Lloyd and McCown 1980.

Rosa media

Compound	Stage I	II
	mg/liter	
Ammonium nitrate	400	400
Boric acid	6.2	6.2
Calcium chloride	96	96
Calcium nitrate	556	556
Cupric sulfate	0.025	0.025
Magnesium sulfate	370	370
Manganese sulfate	22.3	22.3
Potassium phosphate	170	170
Potassium sulfate	990	990
Sodium molybdate	0.025	0.025
Zinc sulfate	8.6	8.6
MS iron	65.1	65.1
Inositol	100	100
Thiamine HCl	1.0	1.0
Nicotinic acid	0.5	0.5
Pyridoxine HCl	0.5	0.5
BA	–	1.0
Glycine	2.0	2.0
Sucrose	20,000	20,000
Agar	(6,000)	6,000
pH 5.7		

Rubus, Blackberry

Blackberries *(Rubus)* respond well to culture with 3- to 6-times multiplication every 3 to 4 weeks.

Explant: 5-cm shoot tips.

Treatment: Remove leaves, except for the youngest ones enclosing the terminal bud. Agitate shoot tips in distilled water with a few drops of detergent for one minute. Rinse for 5 minutes each in 4 changes of distilled water. Mix for 5 minutes in 1/20 bleach with 0.1% Tween 20. Rinse briefly in sterile water, followed by 5 minutes each in 3 changes of sterile water. Trim to 2 to 4 cm.

As a new shoot is produced by an explant, subculture the shoot with an attached piece of the original explant until the shoot reaches 2 to 5 cm.

Media: Modified MS with IBA, BA, and GA₃. Stage I may be in agitated liquid or on agar. Use Stage III medium, or root directly from Stage II in fine greenhouse potting mix, milled sphagnum, or Jiffy 7's.

Light: 200 to 400 f.c. from warm-white fluorescent lamps, 16 hours light/8 hours dark.

Temperature: 25°C (77°F).

References: Borgman and Mudge 1986, Donnelly et al. 1985, Kyte and Kyte 1981, Murashige 1974, Murashige and Skoog 1962, Zimmerman and Broome 1980c.

Rubus (blackberry) media

Compound	Stage I & II	III
	mg/liter	
MS salts	4,628	4,628
Inositol	100	100
Thiamine HCl	0.4	0.4
IBA	–	1.0
BA	1.0	–
GA₃	0.5	–
Sucrose	30,000	30,000
Agar	5,000	5,000
Gelrite	3,000	3,000
Charcoal	–	0.6
pH 5.2		

Rubus, Raspberry

Most raspberries *(Rubus)* are difficult to micropropagate. The larger the explant, the sooner it will start growing but the more apt it is to be contaminated. Excised buds may be cleaner than shoots, but they will take longer to grow. In any case, explants take several weeks to react. Transferring should take place every 2 weeks or more frequently. Experimentation and fine tuning with hormone strength, gel strength, and salt concentrations is important to success. Tissue culture is the fastest way to introduce new raspberry cultivars to the trade.

Explant: One-node sections or buds from active new growth (5- to 8-cm shoots).

Treatment: Wash shoots in water with 0.1% Tween 20 for 15 minutes. Stir in 1/10 bleach for 10 minutes. Treatment time should correspond to tenderness of the shoot. Rinse 4 times in sterile distilled water. Trim to single-node sections or excise buds.

Shade and humidity are critical in growing-on these tissue cultured plantlets. Jiffy 7's or light seedling mixes will help the transition period, during which the plantlets must grow new leaves before they are able to flourish in a normal greenhouse environment.

Media: The multiplication medium is the same as given above for blackberry. For *in vitro* rooting in Stage III, a modified form of Anderson's medium is recommended.

Light: 100 to 300 f.c. from cool-white fluorescent light, 16 hours light/8 hours dark.

Temperature: 25°C (77°F).

References: Anderson 1980, Borgman and Mudge 1986, Donnelly et al. 1985, Gebhardt 1985, Kyte and Kyte 1981, Murashige 1974, Murashige and Skoog 1962, Zimmerman and Broome 1980c.

Rubus (raspberry) media

Compound	Stage I & II	III
	mg/liter	
MS salts	4,628	–
Ammonium nitrate	–	400
Calcium chloride	–	440
Magnesium sulfate	–	370
Potassium nitrate	–	480
Sodium phosphate	–	380
MS minor salts	–	32.8
MS iron	–	65.1
Inositol	100	100
Thiamine HCl	0.4	0.4
IBA	–	1.0
BA	1.0	–
GA$_3$	0.5	–
Sucrose	30,000	30,000
Agar	8,000	8,000
Gelrite	3,000	3,000
Charcoal	–	0.6
pH 5.3		

SOLANACEAE, Nightshade Family

Solanum tuberosum, Potato

Because potatoes *(Solanum tuberosum)* are very subject to disease, tissue culture is an ideal way to multiply disease-free stock. The best method of micropropagation is by nodal propagation. Establishment of the primary explant requires approximately 8 weeks. Nodal multiplication rates vary from variety to variety and from line to line (a line is all those plants derived from a single tuber). With the particular medium given here, roots will develop along with shoots, and

therefore a separate rooting medium is not needed. Transfer to soil is readily accomplished in a mist bench or a humidity tent.

Explant: Apical and axillary buds from sprouted tubers.

Treatment: Dip shoots in 70% alcohol. Stir in 1/10 bleach plus 0.1% Tween 20 for 10 minutes. Rinse 3 times in sterile distilled water. Be sure to provide air to the cultures.

Media: MS salts with inositol, thiamine, kinetin, and GA$_3$.

Light: 300 f.c. from cool-white fluorescent light, 16 hours light/8 hours dark.

Temperature: 23°C (73°F).

References: Goodwin and Adisarwanto 1980, Morel and Muller 1964, Murashige 1974, Murashige and Skoog 1962, Upham 1983.

Solanum medium

Compound	Stage I, II & III
	mg/liter
MS salts	4,628
Inositol	100
Thiamine HCl	0.4
Kinetin	0.5
GA$_3$	0.1
Sucrose	30,000
Agar	7,000
pH 5.7	

VITACEAE, Grape or Vine Family

Vitis vinifera, Grape

The urgent need for virus-free, crown gall–free grapevines is beginning to be met by way of tissue culture propagation of *Vitis vinifera.* Using mere 2-mm explants, plants can be rendered virtually disease free for vineyards or for stocks held in culture. For plants that are already disease-free, larger explants are advised for quicker establishment. By conventional methods, eradicating disease from grapevines is extremely time consuming and costly, and several more years are required to propagate a few hundred plants from a "clean" plant, or 2 years to produce field-ready plants by grafting. In contrast, hundreds of tissue cultured plants can be field ready within a year, either on their own roots or by grafting.

Although prospects look bright for grape micropropagation, the genus *Vitis* remains difficult to manage because of inconsistent responses. Frequent transfer (2-week intervals), special concern for detail, and careful observation with prompt action are particularly crucial for this important plant.

Explant: 3-node shoot tips.

Treatment: Cut 5-cm shoot tips. Remove and discard the expanded leaves. Stir tips in 70% alcohol for one minute or in 7% bleach plus 0.1% Tween for 20 minutes. Rinse 4 times in sterile distilled water. In the transfer hood, cut 3-node tips, and place on soft Gelrite and agar.

Tall jars are preferable to baby-food jars in order to allow enough space for micro-nodal cuttings for transfer. Cuttings may go directly from Stage II to soil for rooting, but they may be quicker to establish if put through Stage III. For Stage IV, use peat/vermiculite (1/4) soil mix or coarse perlite. Either way, water the plantlets with Stage III medium without sucrose. Place in high humidity for 3 weeks, then continue with intermittent mist, gradually reducing humidity.

Media: Use ¾-strength MS with additives.

Light: 300 f.c. from fluorescent light, 16 hours light/8 hours dark.

Temperature: 23 to 30°C (73 to 86°F) is preferred, but not essential for rooting.

References: Barlass and Skene 1978, Briggs 1984, Chee et al. 1984, Harris and Stevenson 1982, Krul and Myerson 1980, Monette 1983, Murashige 1974, Murashige and Skoog 1962, Smith et al. 1992.

Vitis media

Compound	Stage I	II	III
		mg/liter	
MS salts	3,471	3,471	3,471
Sodium phosphate	–	170	150
Adenine sulfate	–	80	–
Inositol	100	100	25
Thiamine HCl	0.4	0.4	0.4
IAA	–	–	0.1
BA	0.1	2.0	–
Sucrose	30,000	30,000	10,000
Agar	500	500	500
Gelrite	1,000	2,000	1,000
pH 5.3			

FORMULA COMPARISON CHART

(amounts are given in milligrams per liter)

	Anderson (1978)	Gamborg et al. (1976)	Gautheret (1942)	Heller (1953)	Hildebrandt, Riker, & Duggan (1946)	Hoagland (1950)	Knop (1865)	Knudson (1946)
Ammonium nitrate (NH_4NO_3)	400	–	–	–	–	–	–	–
Ammonium phosphate ($NH_4H_2PO_4$)	–	–	–	–	–	115	–	–
Ammonium sulfate ($[NH_4]_2SO_4$)	–	134	–	125	–	–	–	500
Boric acid (H_3BO_3)	6.2	3.0	0.05	1.0	3.0	2.86	–	–
Calcium chloride ($CaCl_2 \cdot 2H_2O$)	440	150	–	75	–	–	–	–
Calcium nitrate ($Ca[NO_3]_2 \cdot 4H_2O$)	–	–	500	–	800	945	500–800	1,000
Calcium phosphate, tribasic ($Ca_{10}[PO_4]_6[OH]_2$)	–	–	–	–	–	–	–	–
Cobalt chloride ($CoCl_2 \cdot 6H_2O$)	0.025	0.025	0.05	–	–	–	–	–
Cupric sulfate ($CuSO_4 \cdot 5H_2O$)	0.025	0.025	0.05	0.03	–	–	–	–
Magnesium sulfate ($MgSO_4 \cdot 7H_2O$)	370	250	125	250	720	250	125–200	250
Manganese chloride ($MnCl_2 \cdot 4H_2O$)	–	–	–	–	–	1.81	–	–
Manganese sulfate, monohydrate ($MnSO_4 \cdot H_2O$)	16.9	10	3.0	–	–	–	–	–
Manganese sulfate ($MnSO_4 \cdot 4H_2O$)	–	–	–	0.1	4.5	–	–	7.5
Potassium chloride (KCl)	–	–	–	750	130	–	–	–
Potassium iodide (KI)	–	0.75	0.5	0.1	0.375	–	–	–
Potassium nitrate (KNO_3)	480	2,500	125	–	160	607	125–200	–
Potassium phosphate (KH_2PO_4)	–	–	125	–	–	–	125–200	250
Potassium sulfate (K_2SO_4)	–	–	–	–	–	–	–	–
Sodium molybdate ($Na_2MoO_4 \cdot 2H_2O$)	0.25	0.25	–	–	–	–	–	–
Sodium nitrate ($NaNO_3$)	–	–	600	600	–	–	–	–
Sodium phosphate ($NaH_2PO_4 \cdot H_2O$)	380	150	125	125	132	–	–	–
Sodium sulfate (Na_2SO_4)	–	–	–	–	100	–	–	–
Zinc sulfate ($ZnSO_4 \cdot 7H_2O$)	8.6	2.0	0.18	1.0	3.0	–	–	–
Ferric sulfate ($Fe_2[SO_4]_3$)	–	–	–	50	–	–	–	–
Ferric tartrate ($Fe_2[C_4H_4O_6]_3$)	–	–	–	–	5.0	–	–	–
Ferrous sulfate ($FeSO_4 \cdot 7H_2O$)	55.7	27.8	–	–	–	5.0	–	25
Na_2EDTA	74.5	37.3	–	–	–	–	–	–
Adenine sulfate	80	–	–	–	–	–	–	–
d–Biotin	–	–	–	–	–	–	–	–
Inositol	100	100	–	–	–	–	–	–
Nicotinic acid	–	1.0	–	–	–	–	–	–
Pyridoxine HCl	–	1.0	–	–	–	–	–	–
Thiamine HCl	0.4	10	–	–	–	–	–	–
2iP	5.0	–	–	–	–	–	–	–
IAA	1.0	–	–	–	–	–	–	–
Kinetin	–	–	–	–	–	–	–	–
Glycine	–	–	–	–	–	–	–	–
Sucrose	30,000	30,000	–	–	–	–	–	–
Agar	6,000	–	–	–	–	–	–	–

Notes:

1. These are well-known basic formulas to which are added micronutrients, vitamins, growth regulators, and other compounds, as desired.

Linsmaier & Skoog (1965)	Lloyd & McCown (WPM) (1980)	Morel & Muller (1964)	Murashige & Skoog (1962)	Nitsch & Nitsch (1969)	Schenk & Hildebrandt (1972)	Vacin & Went (1949)	White (1963)
1,650	400	–	1,650	720	–	–	–
–	–	–	–	–	300	–	–
–	–	1,000	–	–	–	500	–
6.2	6.2	–	6.2	10	5.0	–	1.5
400	96	–	440	166	200	–	–
–	556	500	–	–	–	–	300
–	–	–	–	–	–	200	–
0.025	–	–	0.025	–	0.1	–	–
0.025	0.025	–	0.025	0.025	0.2	–	–
370	370	125	370	185	400	250	720
–	–	–	–	–	–	–	7.0
–	22.3	–	16.9	–	10	–	–
22.3	–	–	–	25	–	7.5	–
–	–	1,000	–	–	–	–	65
0.33	–	–	0.83	–	1.0	–	0.75
1,900	–	–	1,900	950	2,500	525	80
170	170	125	170	68	–	250	–
–	990	–	–	–	–	–	–
0.25	0.025	–	0.25	0.25	0.1	–	–
–	–	–	–	–	–	–	–
–	–	–	–	–	–	–	16.5
–	–	–	–	–	–	–	200
8.6	8.6	–	8.6	10	1.0	–	3.0
–	–	–	–	–	–	–	2.5
–	–	–	–	–	–	28	–
2.8	27.8	–	27.8	27.8	15	–	–
37.3	37.3	–	37.3	37.3	20	–	–
–	–	–	–	–	–	–	–
–	–	–	–	0.05	–	–	–
100	100	–	100	100	1000	–	–
–	0.5	–	0.5	5.0	5.0	–	0.5
–	0.5	–	0.5	0.5	0.5	–	0.1
–	1.0	–	0.4	0.5	5.0	–	0.1
–	1.0	–	–	–	–	–	–
1.3	–	–	1.3	0.1	–	–	–
0.001–10	–	–	0.04–10	–	–	–	–
–	2.0	–	–	2.0	–	–	3.0
30,000	20,000	–	30,000	20,000	–	–	20,000
10,000	6,000	–	8,000	8,000	–	–	–

2. Omitted are compounds of beryllium, titanium, nickle, and folic acid. These are specified in some original formulas but are very rarely used.

Appendix A: Metric Conversions

Because some equipment specifications and all media formulas are expressed in metric terms, it is important to be able to read and understand the metric system. Familiarity with metric measurements comes quickly and easily with practice and makes media calculations relatively simple. Length, volume, weight, and temperature are the four basic properties for which measurements are taken. In the English system these are measured in units that are based on different multiples—12 inches in a foot, 4 quarts per gallon, 16 ounces in a pound, and so on. In the metric system, however, all are measured in units based on multiples of 10. The prefixes of metric measurements indicate the multiple. The most common prefixes for metric measurements are

kilo (k)	$= 10^3$	$= 1,000$
centi (c)	$= 10^{-2}$	$= 0.01$
milli (m)	$= 10^{-3}$	$= 0.001$
micro (μ)	$= 10^{-6}$	$= 0.000001$
nano (n)	$= 10^{-9}$	$= 0.000000001$

In the metric system, length is measured in meters (m), volume in liters (l), weight in grams (g), and temperature in degrees centigrade (°C).

Unit	Symbol	Equivalent
LENGTH		
kilometer	km	1000 m (0.62 miles)
meter	m	100 cm (3.3 feet)
centimeter	cm	10^{-2} m (2.5 cm = 1 inch)
millimeter	mm	10^{-3} m
nanometer	nm	10^{-9} m
angstrom	A	10^{-10} m
VOLUME		
liter	l	1000 ml (1.1 quarts; 3.8 liters = 1 gallon)
milliliter	ml	10^{-3} liter
microliter	μl	10^{-6} liter
WEIGHT		
kilogram	kg	1000 g (2.2 pounds)
gram	g	1000 mg (0.035 ounces; 454 g = 1 pound)
milligram	mg	10^{-3} g
microgram	μg	10^{-6} g
nanogram	ng	10^{-9} g
picogram	pg	10^{-12} g

TEMPERATURE

$^\circ$C $\quad = 5/9 \times (^\circ$F$-32)$

$^\circ$F $\quad = (9/5 \times\, ^\circ$C$) + 32$

0°C $\quad = 32^\circ$F, water freezes

22°C $\quad = 72^\circ$F, "room temperature"

37°C $\quad = 98.6^\circ$F, body temperature

100°C $\quad = 212^\circ$F, water boils

Appendix B: The Microscope—Introduction, Use, and Care

A microscope is often necessary to allow you to visualize particles too small to be seen with the naked eye. This is particularly important when dealing with microbial contaminants because most bacteria, yeast, and fungal spores range in size from 2 to 10 μm. The limit of visibility with the naked eye is about 100 μm (0.1 mm).

When properly illuminated, a microscope is able to provide both magnification and resolution. The value of a microscope depends not so much on the amount of magnification, but more on the ability of a microscope to allow the viewer to clearly separate and distinguish (i.e., resolve) two objects that are in close proximity to one another. By definition, *resolution* is the minimum distance that can exist between two points in order that the points can be observed as separate particles. Due to its better resolution, the simple microscope invented by Antoni van Leeuwenhoek was superior to Robert Hooke's compound microscope even though Hooke's provided greater magnification. The lenses employed by Hooke were optically flawed, which caused the images to blur. We now know that objective lenses with chromatic (color) and spherical distortions will diminish resolution. Thus, the best image is not the one that is largest but the one that is clearest.

The optical and mechanical features of a microscope are illustrated in the picture below. The fixed stage of a microscope has a clamp for holding a glass specimen slide on the surface of the stage and which allows movement of the slide around the surface of the stage. The movement is actuated by 2 stage adjustment knobs located below the stage, one for horizontal movement and the other for vertical movement. Coarse and fine focus adjustment knobs bring the slide specimen into focus. The 2 main operating systems of a compound microscope are optics and illumination.

Optical system

The major characteristics of the 3 most commonly used objectives are:

Objective	Low	High Dry	Oil
	16 mm	4 mm	1.8 mm
Magnification*	10×	45×	96×
Working distance (mm)	7.0	0.6	0.15
Iris diaphragm opening	Least	Some	Greatest
Numerical aperture (N.A.)*	0.25	0.66	1.25

*These figures are inscribed on the outside surface of each objective.

From the above table you can see that the magnification of the oil immersion objective is almost 10 times greater than the magnification of the low-power objective. The total magnification is the product of the magnification of the ocular lens (usually 10×) and the magnification of the objective lens. For example, a 45× objective used with a 10× ocular magnifies the object on the slide 450 diameters—a diameter is the unit of measure used to describe the magnifying power of a lens.

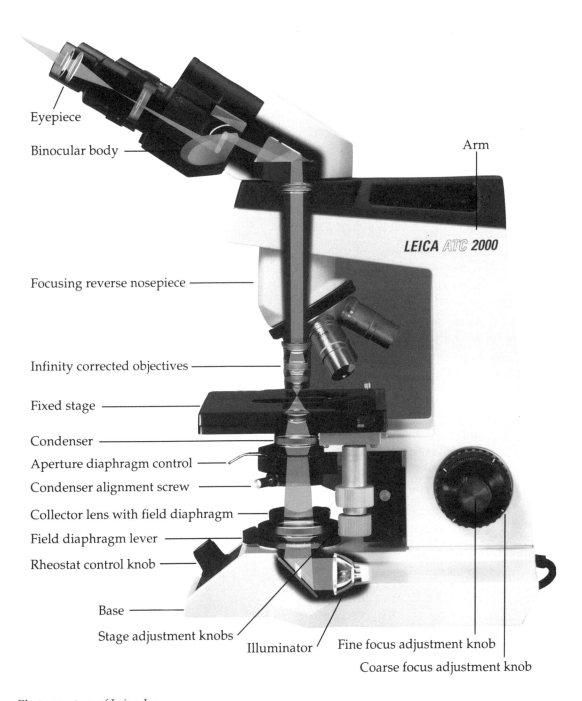

Eyepiece

Binocular body

Arm

LEICA ATC 2000

Focusing reverse nosepiece

Infinity corrected objectives

Fixed stage

Condenser

Aperture diaphragm control

Condenser alignment screw

Collector lens with field diaphragm

Field diaphragm lever

Rheostat control knob

Base

Stage adjustment knobs

Illuminator

Fine focus adjustment knob

Coarse focus adjustment knob

Photo courtesy of Leica, Inc.

The working distance is the distance, in millimeters, from the tip of the objective lens, when in focus, to the principal point of focus. You can see that the working distance of the oil immersion lens is dangerously close to the coverslip surface.

The iris diaphragm on a microscope is similar to the iris of the eye, which opens in diminished light and closes with excess light. With respect to the microscope, the objective lens being used determines how much to open or close the iris diaphragm. The oil immersion objective requires the largest iris diaphragm opening because the bottom tip of the lens is very small and admits little light. The light is also conserved by using a drop of lens immersion oil on the surface of the coverslip. The lens immersion oil deflects into the lens light rays that otherwise would be diffracted (bent outwards from the lens opening) in the absence of the oil.

The numerical aperture relates directly to the resolving power of the objective, but not to magnification. The oil immersion objective has a resolving power 5 times greater than that of the low-power objective.

Illumination system

The best resolution and magnification on a microscope can only be attained with optimal illumination. Centering the light on the specimen is especially important. On less-expensive microscopes, the centering is provided. In microscopes used for more research-oriented purposes, the proper centering of light becomes critical. This is achieved by using the condenser alignment screw, shown in the diagram above.

The field diaphragm lever adjusts the amount of light entering the condenser system. Move the condenser knob (not shown) up and down to achieve the best focus. The amount of light entering the condenser system can be further adjusted with the condenser aperture diaphragm control. The amount of light emitted by the condenser should be tailored to the size of the microscope objective lens. Be cautioned: too much light results in glare from stray light, which reduces resolution.

MICROSCOPE OPERATING PROCEDURE

1. Plug in the microscope light source or illuminator. If you have a voltage transformer, adjust it to 3 volts for the low-power objective and increase to 6 volts for the higher power objectives.
2. Cover a stained slide with a coverslip. Place the slide, *specimen side up*, in the stage slide holder.
3. Using the stage adjustment knobs, position the specimen over the center of the light coming from the condenser.
4. Raise the objective nosepiece using the coarse focusing knob and rotate the low-power objective (10×) into place with the nosepiece.
5. While observing from the side, use the coarse focus adjustment knob to lower the low-power objective as far as it will go.
6. Using the condenser knob, lower the condenser about 0.1 mm (approximately paper thickness) below the highest position.
7. While looking through the ocular, raise the objective using the coarse focus adjustment knob until what appears as a blur comes in the microscope field. Complete the focusing process using the fine focus adjustment knob.
   If you are using a binocular microscope, focus first with one eye only, then close that eye and focus the specimen with the other eye using the ocular adjustment located on the eyepiece. While looking through the oculars with both eyes, turn the knob located between the oculars to adjust the distance between the two

oculars so that they agree with the distance between your eyes. High-eyepoint oculars are available for people who wear glasses.

8. For final light adjustment, remove one of the oculars and look through the ocular tube. Adjust the condenser aperture diaphragm control until the edges of the diaphragm just disappear from view. This adjustment, often neglected, offers the best resolution.

9. Look around for an area you wish to study more intensively. Switch to the next higher objective (high dry, 45×). If your microscope is parfocal (the objectives can be changed without changing the focus), either the specimen will already be in focus or it will almost be in focus, requiring only fine focus adjustment. If using a microscope that is not parfocal, you will need to repeat steps 4, 5, and 7 for the higher objective. For optimum resolution, step 8 should be repeated for each objective.

10. The highest magnification available for viewing a specimen is the oil immersion objective (100×). First rotate the high dry objective to one side so that a drop of lens immersion oil can be placed on the coverslip.

11. If your objectives are parfocal, rotate the oil immersion objective lens into the oil. If done carefully, the objective lens will not come into contact with the coverslip surface. Use only the fine focus adjustment knob, taking care not to hit the slide. If your objectives are not parfocal, it will be necessary to look from the side—not through the ocular—while carefully lowering the oil immersion objective into the oil using the coarse focus adjustment knob. Now looking through the ocular, use the fine focus adjustment knob for final focusing. Newer microscopes usually have spring-loaded objectives and/or nosepieces that help prevent damage to the lens when the objective touches the slide. Nevertheless, great care is required.

12. When you are finished observing the specimen, wipe excess oil from the oil immersion objective lens with lens paper and, if necessary, clean the ocular with lens paper moistened with distilled water. This is especially important in hot weather when sweat from your eyes, acidic in nature, contacts the ocular lens. The presence of other foreign particles on the ocular can be determined by rotating the ocular lens manually while looking through the microscope. Any dirt on the ocular lens also will rotate as you rotate the lens, thus indicating the presence of dirt. If there is dirt on the ocular, clean both the upper and lower surfaces of the ocular lens. If any dirt remains after cleaning, that is usually evidence either of a scratched lens surface or of dirt particles on the inside surface of the lens.

MICROSCOPE CARE

- Always remove excess oil from the oil immersion lens before storing.
- Before storing the microscope, rotate the nosepiece to the lowest objective and raise it to the highest position with the coarse adjustment knob.
- Use a dust cover and a box for storage.
- Always transport the microscope using both hands, one holding the arm of the microscope and the other under the base.
- Do not force microscope instrument knobs. Consult a specialist about any problems.
- Do not attempt to clean the *inside* of objective lenses. This also requires a specialist.

Appendix C: Professional Organizations

International Association for Plant Tissue Culture
Department of Plant Sciences
Texas A & M University
College Station, Texas 77843, U.S.A.
Dr. Roberta H. Smith, National (U.S.) Correspondent
Publications: Newsletter; *Proceedings of the International Congress of Plant Tissue and Cell Culture*

Tissue Culture Association, Inc.
8815 Centre Drive
Suite 210
Columbia, Maryland 21045, U.S.A.
410-992-0946
Publications: *In Vitro*; *TCA Report*; *Plant Cell, Tissue, and Organ Culture*

American Society for Horticultural Science
701 North Saint Asaph Street
Alexandria, Virginia 22314, U.S.A.
Publications: *Journal of the American Society for Horticultural Science*; *HortScience*; *HortTechnology*

The International Plant Propagators' Society, Inc.
Dr. John Wott
Center for Urban Horticulture, GF15
University of Washington
Seattle, Washington 98195, U.S.A.
Publications: *The Plant Propagator*; *Proceedings of the International Plant Propagators' Society* (Annual)

Appendix D: Suppliers

Agdia, Inc.
30380 County Road 6
Elkhart, Indiana 46514, U.S.A.
219-264-2014; 800-622-4342
ELISA test kits for plant pathogens

Aldrich Chemical Co.
940 West Saint Paul Avenue
Milwaukee, Wisconsin 52333, U.S.A.
414-273-3850; 800-558-9160
Chemicals

American Biotechnology Laboratory
P.O. Box 3222
Lowell, Massachusetts 01859, U.S.A.
Publish a free trade magazine and a yearly
    buyer's guide

American Type Culture Collection
12301 Parklawn Drive
Rockville, Maryland 20852, U.S.A.
301-881-2600; 800-638-6597
Plant viruses and antisera (antibodies), bacteria
    and bacteriophages, yeasts, filamentous
    fungi

Barnstead/Thermolyne
2555 Kerper Boulevard
P.O. Box 797
Dubuque, Iowa 52004-0797, U.S.A.
Phone: 319-556-2241; 800-553-0039
Fax: 319-556-0695
Water purification systems

Baxter Diagnostic Inc.
    Scientific Products Division
1 Baxter Parkway
Deerfield, Illinois 60015, U.S.A.
708-270-2001
Worldwide source of laboratory equipment
    and supplies

Becton Dickinson Microbiology Systems
P.O. Box 243
Cockeysville, Maryland 21030-0243, U.S.A.
410-584-7154; 800-638-8663
Microbiological media; publish *The Manual of
    BBL Products and Laboratory Procedures*

Better Plastics, Inc.
2206 North Main Street
Kissimmee, Florida 34744, U.S.A.
Phone: 407-846-3127; 800-932-7151
Fax: 407-846-2584
Autoclavable, polypropylene, 16-oz culture
    containers and lids

Brinkman Instruments Co.
#1 Cantiague Road
Westbury, New York 11590, U.S.A.
516-334-7500; 800-645-3050
Micropipets, instruments

Cadillac Plastic and Chemical Co.
18292 Andover Park West
Seattle, Washington 98188, U.S.A.
206-682-7252
Polycarbonate sheets

Calbiochem Behring
    Division of American Hoechst Corp.
P.O. Box 12087
San Diego, California 92112, U.S.A.
Phone: 619-450-9600; 800-854-3417
Fax: 619-776-0999
Chemicals

Carolina Biological Supply Co.
2700 York Road
Burlington, North Carolina 27215, U.S.A.
910-584-0381; 800-334-5551
or
Box 187
Gladstone, Oregon 97027, U.S.A.
800-547-1733
Tissue culture supplies

Clean Room Filters & Supplies, Inc.
Lehigh Valley Industrial Park, #4
54 Commerce Way, Suite 170
Bethlehem, Pennsylvania 18017, U.S.A.
610-974-8077; 800-334-9142
HEPA filters

Consolidated Plastics Co., Inc.
8181 Darrow Road
Twinsburg, Ohio 44087, U.S.A.
216-425-3900; 800-362-1000
Autoclavable plastic growth jars, clean room
    uniforms, carts, shipping supplies, marking
    tapes

DENT-EQ
101/1, Surveyor Street
Basavanagudi, Bangalore-560 004,
Karnataka, India
Phone: 91-080-626459
Fax: 91-080-610666
Glass bead sterilizers

Difco Laboratories
P.O. Box 331058
Detroit, Michigan 48232, U.S.A.
313-462-8500; 800-521-0851
Microbiological media; publish *Difco Manual*

Edmund Scientific
Edscorp. Building
Barrington, New Jersey 08007-1380, U.S.A.
Phone: 609-547-8880
Fax: 609-573-6295
Science products for educators, students, and
    inventors

Eppendorf Scientific, Inc.
545 Science Drive
Madison, Wisconsin 53711, U.S.A.
608-276-9869; 800-421-9988
Electroporation equipment, micropipetters

Fisher Scientific
711 Forbes Avenue
Pittsburgh, Pennsylvania 15219, U.S.A.
800-766-7000; 800-955-5090 (international)
Equipment and supplies

Fungi Perfecti
P.O. Box 7634
Olympia, Washington 98507, U.S.A.
206-786-1105
Sterile culture and growing room equipment
    and supplies, desktop laminar flow hoods

Gelman Sciences
600 South Wagner Road
Ann Arbor, Michigan 48106-1448, U.S.A.
Phone: 313-665-0651; 800-821-5525
Fax: 313-761-1208
Membrane filters, life sciences equipment;
    publish *Laboratory Filtration*

W. W. Grainger, Inc.
2001 Grand Street
Seattle, Washington 98144, U.S.A.
206-251-5030
Motors

Hemco Corporation
111 North Powell
Independence, Missouri 64056, U.S.A.
816-796-2900; 800-779-4362
Clean rooms, HEPA filters, lab furniture

Integrated Air Systems, Inc.
3750 Cohasset Street
Burbank, California 91504, U.S.A.
213-842-5211
Laminar air flow hoods, HEPA filters

ouan, Inc.
Route 8, Box 1202
P.O. Box 2716
Winchester, Virginia 22601, U.S.A.
703-665-0863; 800-662-7477
Centrifuges

Lab Equipment Magazine
P.O. Box 14000
Dover, New Jersey 07801-9873, U.S.A.
Equipment information, vendors; free
    subscription

Labconco Corp.
3811 Prospect
Kansas City, Missouri 64132-9971, U.S.A.
816-333-8811; 800-821-5525
Water purification systems

LAB-LINE Instruments, Inc.
15th and Bloomingdale Avenues
Melrose Park, Illinois 60160-1491, U.S.A.
Phone: 708-450-2600; 800-525-5463
Fax: 708-450-0943
Tissue culture rotators

Leica, Inc.
111 Deer Lake Road
Deerfield, Illinois 60015, U.S.A.
708-405-0123; 800-248-0123
Microscopes

Liberty Industries
133 Commerce Street
East Berlin, Connecticut 06023, U.S.A.
Phone: 860-828-6361; 800-828-5656
Fax: 860-828-8879
Laminar air flow hoods, HEPA filters

Li-Cor, Inc.
4412 Superior Street
P.O. Box 4425
Lincoln, Nebraska 68504, U.S.A.
402-467-3576
Light meters

Life Technologies, Inc., GIBCO BRL
3175 Staley Road
Grand Island, New York 14072, U.S.A.
800-952-9166
Media

Magenta Corp.
3800 North Milwaukee Avenue
Chicago, Illinois 60641, U.S.A.
312-777-5050
Culture containers

Midway Co.
5875 DW Van Horn Tavern Road
Columbia, Missouri 65203, U.S.A.
800-243-3220
Lee Production lead pots

Millipore Corp.
80 Ashby Road
P.O. Box 9125
Bedford, Massachusetts 01730, U.S.A.
617-275-9200; 800-645-5476
Membrane filters, water analysis equipment,
    0.45 μM disposable filters

Mine Safety Appliances Co.
P.O. Box 426
Pittsburgh, Pennsylvania 15230, U.S.A.
412-967-3000; 800-672-2222
Air sampling pumps and accessories

Monarch Marking
P.O. Box 608
Dayton, Ohio 45401, U.S.A.
513-865-2123; 800-543-6650
Labelers

Nalge Co.
75 Panorama Creek Drive
P.O. Box 20365
Rochester, New York 14602, U.S.A.
716-586-8800
Labware: graduated cylinders, funnels, beakers,
    flasks, etc.

New Brunswick Scientific Co., Inc.
44 Talmadge Road.
Edison, New Jersey 08818, U.S.A.
Phone: 908-287-1200; 800-448-2298
Fax: 908-287-4222
Rotators, shakers, automatic preparation
    equipment

Northern Steel
600 Andover Park East
Seattle, Washington 98188, U.S.A.
206-767-5383; within Washington state: 800-562-
    1086
Slotted angles for shelving supports

Research Organics, Inc.
4353 East 49th Street
Cleveland, Ohio 44125-1083
Phone: 216-883-8025 (international); 800-321-
    0570
Fax: 216-883-1576
Biochemicals, Gellan (Gelrite), agar

Scientific Supply and Equipment, Inc.
926 Poplar Place South
Seattle, Washington 98144, U.S.A.
206-324-8550
Chemicals, supplies, and equipment

Scott Laboratories, Inc.
1137 Janis Street
Carson, California 90746, U.S.A.
213-537-6800
Gelrite (agar substitute)

Harry Sharp & Son
900 Lind Avenue SW
Renton, Washington 98005, U.S.A.
206-235-4510; 800-426-7657
Todd Planter flats, greenhouses, greenhouse
    supplies

Sigma Chemical Co.
P.O. Box 14508
St. Louis, Missouri 63178, U.S.A.
314-771-5750; 800-325-3010
Chemicals, tissue culture media and supplies,
    cell culture, biotech and indexing supplies,
    books

Simon Keller Ltd.
Lyssachstrasse 83
3400 Burgdorf, Switzerland
Glass bead sterilizers

SKC Inc.
863 Valley View Road
Eighty Four, Pennsylvania 15330, U.S.A.
412-941-9701; 800-752-8472
Air sampling pumps and accessories

Spencer-Lemaire Industries, Ltd.
11413 120th Street
Edmonton, Alberta, Canada T5G 2Y3
Phone: 403-451-4318
Fax: 403-452-0920
Planter containers for growing on

Terra Universal, Inc.
700 North Harbor Boulevard
Anaheim, California 92805-2553, U.S.A.
714-526-0100; 800-767-0135
Clean room handling and products, laboratory
    equipment, manuals and books

VWR Scientific
P.O. Box 3551
Seattle, Washington 98124, U.S.A.
206-575-1500; 800-333-6336; 415-468-7150
    (international)
Supplies and equipment

Ward's Natural Science Establishment, Inc.
5100 West Henrietta Road
P.O. Box 72912
Rochester, New York 14692-9012, U.S.A.
716-359-2502; 800-962-2660
Biological supplies, tissue culture supplies

Glossary

abscission layer: a thin-walled layer of cells that forms for the normal separation of leaves or fruit from plants.

adventitious: growing from unusual locations, such as aerial roots from stems; or the formation of buds at locations other than nodes or stem tips.

aeration: the process of mixing, shaking, or venting a culture to provide it with air.

agar: a polysaccharide gel derived from certain red algae. Agar is used for making semi-solid or gelatinous media for growing tissue cultures and microbes.

alkaloid: any of a number of colorless, crystalline, bitter organic substances with alkaline properties. Many are found in plants; they can have toxic effects on humans. Caffeine, quinine, and strychnine are examples of alkaloid substances.

amino acid: a nitrogenous organic compound that serves as a structural unit of proteins.

anther culture: the *in vitro* culture of anthers to obtain haploid plants.

antioxidant: a substance used to prevent phenolic oxidation, the browning of cultures as a result of bleeding phenolic exudates.

apex (pl., apices; adj., apical): tip.

apical dominance: the inhibition of lateral growth as a result of auxins produced by the apical bud.

apical meristem: the new, undifferentiated tissue at the microscopic tip of a shoot or root. The apical meristematic dome of stem tips—which is usually about 0.5 mm or less—along with the accompanying leaf primordia are commonly used as explants.

artificial seed: a somatic embryo that has been coated or encapsulated and grown as true seed.

aseptic: free from microorganisms. The terms "sterile technique" and "aseptic technique" are used somewhat interchangeably.

asexual: without sex, vegetative.

atom: the smallest particle of an element that retains the chemical characteristics of that element.

atomic weight: a system of relative weights of atoms.

autoclave: a vessel for sterilizing using steam under pressure. It is usually distinguished from a pressure cooker by its size and design. Also the verb form, to sterilize in such a vessel.

autotrophic: able to produce one's own food from inorganic substances.

auxins: a group of growth hormones associated with root initiation, lateral bud inhibition, and cell enlargement.

axil: the point of attachment of lateral growth with the stem.

ballast: a type of transformer wired into fluorescent lamps.

biosynthesis: the formation of chemicals by living cells.

bleeding: the purplish black or brown coloration occasionally found in media due to phenolic products given off by explants or transfers.

bridge: a piece of filter paper or other device placed within a test tube of liquid medium to hold the culture out of the liquid. Also known as rafts or floats.

callus: a proliferating mass of disorganized, mostly undifferentiated or undeveloped cells.

callus culture: the multiplication of callus cells in sterile culture.

cambium: the thin layer of tissue between the wood and bark of a stem.

carbohydrate: any of certain organic compounds composed of the elements carbon, hydrogen, and oxygen, including sugars, starches, and cellulose.

carcinogen: any substance that causes cancer.

catalyst: a substance that promotes the rate of a chemical reaction but which is not itself changed chemically.

cell: the basic physical unit of living organisms.

cell culture: the multiplication *in vitro* of single cells or clumps of cells not organized as tissues; often included in the broad term "tissue culture."

cell suspension culture: the culture of single cells or clumps of cells suspended in a liquid medium.

centrifuge: a device for spinning test tubes of liquid cultures for the purpose of separating particles, by density or size, by means of centrifugal force. Also the verb form, to separate particles by means of centrifugal force.

chimera: a plant that contains tissues of more than one genotype, such as certain plants with variegated foliage.

chlorosis: the absence of green pigments in plants due to lack of light, nutrient deficiency, or disease. Chlorotic plants often have a yellowed appearance.

chromosome: the structure in cell nuclei that carries the genetic material contained within genes.

clone: the plants produced asexually from a single plant, and thus of genotype identical to one another and to the parent plant.

coconut milk (coconut water): the liquid endosperm of coconuts.

coenzyme: an organic substance that activates an enzyme or accelerates its action.

coleoptile: a protective membrane surrounding the primary leaf of grass seedlings. It is noted for its auxin content, as demonstrated by Frits Went's classic experiment.

compound: a substance made up of 2 or more elements chemically combined in fixed proportions.

conductivity: a measure of the ability of water to conduct electricity. The conductivity level indicates the concentration of ions dissolved in the water, and thus the purity of the water. It is often measured in units called siemens (S).

contaminant: any unwanted microorganisms that appear in tissue culture media and overgrow, inhibit, poison, or otherwise pollute the desired culture.

cotyledon: the first leaf or seed leaves. Monocots (monocotyledons) have one cotyledon and dicots (dicotyledons) have two.

crown: the often fleshy base of a plant where the stem joins the root.

cultivar: a named plant variety in cultivation.

culture: a plant or plant part growing *in vitro*. Also the verb form, to grow a plant *in vitro*.

cytokinins: a group of growth hormones that induce bud formation and shoot multiplication and promote cell division and enlargement.

deciduous: leaf fall at a particular season or stage of growth, as in trees and shrubs that lose their leaves for the winter, as opposed to evergreen.

dedifferentiate: to revert to a non-specialized or undifferentiated state, as in cells or tissues.

deionize: to remove soluble minerals and certain organic salts from water by passing the water through an ion-exchange unit.

dicotyledon (dicot): a flowering plant that contains two cotyledons (seed leaves). Dicotyledons are members of the class Dicotyledoneae and include species of trees, shrubs, and herbaceous plants.

differentiate: to initiate the growth of new and varied tissues or organs for specialized functions; process is differentiation.

diploid: having two sets ($2n$) of chromosomes, which is typical of vegetative (somatic) cells.

disinfect: to destroy pathogenic microorganisms using chemical agents (disinfectants).

distill: to remove dissolved minerals, particulates, and organic matter from water by boiling the water and then passing the steam and vapor through a condensing coil, from where the recondensed, distilled water is collected in a container.

DMSO (dimethyl sulfoxide): a toxic solvent that is sometimes used in minute quantities to dissolve certain organic compounds for plant tissue culture media. It readily passes through skin.

dormancy: a period in which there is no growth or little physiological activity; a resting stage. Dormancy can sometimes be broken by cold treatment or by applying gibberellic acid (GA_3).

element: a substance composed of atoms of the same type and which cannot be separated into simpler substances by any usual chemical means.

embryo culture: the *in vitro* culture of embryos excised from seeds, or embryos induced to form from somatic cells (*see* **somatic embryogenesis**).

embryogenesis: the formation of an embryo, either in the seed (sexual) or in culture (asexual).

embryoid: an embryo-like somatic structure that develops in some cell and callus cultures and is capable of developing into an embryo and, subsequently, a whole plant.

endosperm: the nutritive tissue or liquid surrounding a developing embryo in a seed.

enzyme: a specialized protein that acts as an organic catalyst in specific chemical reactions.

epinasty: an abnormal downward bending of a leaf due to the more rapid growth of the upper surface. It may be an indication of a nutritional deficiency or imbalance.

etiolate: to develop long, spindly growth with low chlorophyll. Etiolation results from insufficient or lack of light.

excise: to remove by cutting.

explant: the part of a plant used to start a culture.

exudate: a substance that bleeds (exudes) into a medium from a culture, often brown or purplish black lethal phenolics.

flaming: the procedure of sterile technique whereby an instrument is dipped in alcohol and then set on fire to burn off the alcohol and further sterilize the instrument.

foot-candle (f.c.): a unit of measure of light intensity, based on the illumination on one square foot of surface that is one foot away from a standard candle. One f.c. is roughly equivalent to 10 lux (0.093 f.c. = 1 lux). This unit is obsolete in academia, having been replaced by photosynthetically active radiation (PAR) units such as micromoles per square meter per second ($\mu mol\ m^{-2}\ sec^{-1}$), microeinsteins per square meter per second ($\mu E\ m^{-2}\ sec^{-1}$), or watts per square meter (wm^{-2}). The unit foot-candle is retained in this work for its simplicity and because meters reading in foot-candles are inexpensive and readily available.

formula: (1) a chemical compound expressed in letters, numbers, or symbols to indicate its composition; (2) the ingredients, or recipe, for tissue culture media.

gamete: the male or female reproductive cell; also known as germ cells and as microspore or pollen (male) and megaspore or ovule (female). Gametes have a haploid supply of chromosomes (n) and are the product of meiosis.

gene: a segment of base pairs in a DNA molecule that codes for particular protein or proteins determining hereditary characteristics of the organism.

genetic engineering: the manipulation of genes and chromosomes to vary genetic characteristics.

genetic instability: the tendency of cells to mutate.

genetic variation: any variation in characteristics as a result of genetic inheritance or mutation.

genotype: the genetic make-up of an organism.

germ cell: *see* **gamete**.

gibberellins: a group of growth regulators that influence cell enlargement.

growth regulator: any organic compound that influences growth and multiplication, such as cytokinins and auxins. *See also* **hormones**.

habituated: a state wherein a culture no longer responds to a substance, such as a hormone or nutrient, because the culture no longer requires that hormone or nutrient or is unable to utilize it.

haploid: the number of chromosomes (*n*) in reproductive or germ cells, which is half the normal number of chromosomes (2*n*) in vegetative cells.

hardening off: the process of gradually acclimating plants to tougher environmental conditions, most notably the less-humid conditions of the field or greenhouse than is experienced in culture or during rooting.

HEPA: high efficiency particulate air filters, an essential component of laminar air flow hoods to help ensure sterile transfer work.

herbaceous plant: a seed plant that lacks woody tissues.

heterotrophic: requiring external organic substances as sources of food and nutrients.

hood: *see* **transfer chamber**.

hormone: any naturally occurring or synthetic chemical, such as cytokinins and auxins, that affects growth and development. *See also* **growth regulator**.

hot plate/stirrer: an electrical device that serves as both a hot plate and a magnetic stirrer; the heating and stirring features can be used simultaneously or independently of one another.

hybrid: a plant created from a cross between two genetically different plants by means of sexual reproduction, or artificially by protoplast fusion and other methods of gene transfer in genetic engineering.

hydrate: a compound with chemically bound water. *See* **water of hydration**.

hypocotyl: the portion of a seedling stem below the cotyledons and above the roots.

incubate: to grow under favorable conditions, usually in an incubator or other cabinet providing controlled environmental conditions.

indexing: the testing of plants for the presence of pathogens or contaminants.

indicator plant: a plant that readily expresses the symptoms of a particular disease, onto which a part of another plant is grafted to determine if the latter plant has the suspect disease.

inorganic chemical: not organic, usually lacking carbon.

intergeneric: said of a cross between two or more different genera.

internode: the section of stem between two nodes.

interspecific: said of a cross between two or more different species.

in vitro: literally, *in glass* (Latin). Often used interchangeably with the terms "tissue culture" and "micropropagation," it refers to growing under artificial conditions.

in vivo: literally, *in life* (Latin). Occurring naturally or under natural conditions; not *in vitro*.

ion: any electrically charged atom or group of atoms, whether positively or negatively charged. When a compound chemically separates into negative and positive ions, the process is called ionization.

juvenility: a state of growth associated with the young or immature stage in a plant's life cycle prior to its ability to flower.

laminar air flow: the even flow of air in a specific, layered direction and without turbulence. *See* **transfer chamber**.

layering: a form of vegetative reproduction in which a shoot or twig comes in contact with the soil, forms roots, and grows on as a new plant.

macerate: to separate cells by cutting, grinding, soaking, or other physical or chemical action.

magnetic stirrer: a stirring device in which a whirling magnet attracts a magnetized stir bar placed inside a glass container on the plate. It is often combined with a hot plate to allow for heat while stirring.

medium (pl., media): (1) the substrate for plant growth, specifically the mixture of certain chemical compounds to form a nutrient-rich gel or liquid in or on which tissue cultures are grown; (2) a soil mix.

meiosis: the reduction division process by which haploid (*n*) sex cells (gametes) are produced. It is a reduction division because the number of chromosomes is reduced from 2*n* to *n*.

meniscus: the curved upper surface of a column of liquid.

meristem: the undifferentiated cells at the tips of roots and shoots, in stems as cambium, and in other locations on a plant. Meristematic tissue is the basis of plant growth and development. The term is used loosely in this book to denote a microscopic shoot tip, usually under 1.5 mm and containing one or two leaf primordia, used as an explant. Also the verb form, to excise such small shoot or root tips.

meristem culture: the *in vitro* culture of meristematic tissue; also misused more broadly to denote micropropagation.

mesophyll: the thin, soft tissue between the upper and lower epidermis of the leaf.

metabolism: the physiological activities of an organism.

metabolite: a product of metabolism. *See also* **secondary product**.

micron: an obsolete term for one one-thousandth of a millimeter (0.001 mm), now called a micrometer (μm). The symbol for a micron is μ.

micropropagation: literally, propagation on a very small scale; vegetative multiplication *in vitro*. It is used interchangeably with the terms "tissue culture" or "*in vitro* culture."

mites: any nearly microscopic animals of the order Acarina, class Arachnida. They can be a devastating contaminant of plant tissue cultures.

mitosis: the process of vegetative cell division in the which chromosomes duplicate and then separate.

molar solution: one mole (one gram molecular weight) of a substance in one liter of solution.

mole: the molecular weight of a compound expressed in grams.

molecular weight: the sum of atomic weights in a molecule.

molecule: the smallest quantity into which a chemical compound can be divided and still retain the characteristic properties of that compound.

monocotyledon (monocot): a flowering plant that contains a single cotyledon (seed leaf). Monocotyledons are members of the class Monocotyledoneae. Examples of monocots include grasses, bulbs, and orchids.

morphogenesis: the anatomical and physiological development of an organism, the study of which is called morphology.

mutagen: a mutant-inducing agent, such as radiation, ultraviolet light, or carcinogenic chemicals.

mutagenesis: the formation of mutants.

mutant: a plant that exhibits some variation (mutation) in characteristics due to a rearrangement or other change in genetic make-up.

mutation: a sudden and permanent change in the genetic order of an individual, due to a change in genes or chromosomes or chromosome number, that will alter some characteristic.

mycorrhiza: a fungus that associates, usually symbiotically, with plant roots.

node: the site on a stem where a bud or leaf is attached.

nucleic acid: any of a group of complex organic acids found especially in the nucleus of all living cells as DNA and RNA.

nucleotide: the building blocks of DNA and RNA, each containing a nitrogenous base and its contingent sugar and phosphate.

nurse culture: a culture technique whereby a filter paper separates a callus culture (the nurse culture) from a cell culture, which in turn receives nutrients and growth-promoting factors diffused through the paper from the callus and from the medium.

organic chemical: any of the compounds made up of molecules containing at least one carbon atom.

organogenesis: the formation of organs, such as leaves, shoots, or roots, from cells or tissues.

osmosis: the diffusion of a fluid through a membrane from one solution into a solution of higher density or ion concentration. **Reverse osmosis**, the forcing of a solution through a membrane in the opposite direction, is often used for purifying water.

osmotic pressure: the pressure potential existing in the diffusion of a solvent through a semipermeable membrane into a more concentrated solution in order to equalize the concentrations on both sides of the membrane.

ovule: the structure within a flower that develops into a seed after fertilization, consisting of the egg cell, embryo sac, nucellus, and integuments (future seed coat). Ovules are sometimes used as explants and may be fertilized *in vitro*.

Parafilm: a stretchable, breathable, easily cut adhesive tape used to seal culture containers; it is readily available from scientific equipment supply companies.

parenchyma: a large, thin-walled cell that has the capacity to regenerate and differentiate. They are the most common type of plant cell.

pathogen: a disease-causing microorganism.

pedicel: a stem or stalk of a flower.

petiole: a stem or stalk of a leaf.

pH: a measure of the degree of acidity or alkalinity of a solution as indicated by the concentration of hydrogen ions. It is measured on a scale of 1 to 14, with levels below 7 indicating acidity and above 7 indicating alkalinity.

phenolics: certain organic metabolic by-products. Appearing as purplish black or brown substances, they sometimes exude from new cultures into the surrounding medium and can be toxic to the culture.

photoperiod: the length of time of exposure to light.

pipet: a slender glass tube used to suction small amounts of liquid for the purpose of measuring and transferring such small amounts. Also the verb form, to use such an instrument to transfer small liquid amounts.

plagiotropic: growth that is horizontal rather than vertical, usually referring to conifers in which the main stem does not grow upright.

plasmolysis: the separation of cytoplasm from the cell wall by the removal of water from the protoplast.

plate: to spread a thin film of cells or single cells on an agar medium, usually in a petri dish (petri plate).

pollen culture: the culture of pollen grains, or the anthers containing pollen grains, to produce haploid plants.

polyploid: a multiple of the basic number of chromosomes (*n*), such as triploid (3*n*), tetraploid (4*n*), etc.

polysaccharide: a carbohydrate made up of a chain of one or more simple sugars linked together in its chemical structure.

precipitate: a solid substance that separates out of media solutions and renders those precipitated elements unavailable as nutrients. Also the verb form, the process of such separation.

premix: media formulas prepared and sold commercially in (usually) a dry form.

primordium (pl., primordia): a plant organ in its earliest stage of differentiation and development.

propagule: a small bit of plant that is being propagated; a transfer.

proteins: a large group of complex organic substances made up of amino acids and other diverse elements. They may serve enzymatic or structural functions.

protoplast: a cell that lacks a cell wall but has a membrane.

protoplast fusion: the uniting of two protoplasts.

reduction division: *see* **meiosis**.

regeneration: the formation of new plants or plant parts.

resin bed: with respect to water purification, a tank or cartridge containing charged resin particles to attract ions of opposite charge from impure water.

resistivity: a measure of the resistance of water to the flow of an electrical current. Measured in units of ohm-centimeters (ohm-cm), it is the reciprocal of conductivity.

resolution: the ability of a microscope lens system to give individuality to 2 particles that lie in close proximity to one another.

rotator: a device that rotates culture containers, thus agitating the contents of cultures in liquid media.

scape: a leafless flower stalk growing from the root crown.

secondary product: a product of plant metabolism that is not primarily related to growth and reproduction, such as medicinals, flavorings, dyes, pesticides, etc.

senescence: aging.

somatic: vegetative, non-sexual. All plant cells other than reproductive cells are somatic.

somatic embryogenesis: the formation of embryos from somatic cells.

somatic hybrid: a vegetatively derived hybrid, as by protoplast or cell fusion.

sport: a plant or plant part that has undergone a mutation.

stages of culture: Stage I, establishment; Stage II, multiplication; Stage III, rooting; Stage IV, acclimatization.

sterile technique: the protocol for aseptically transferring cultures. Also called aseptic technique.

stock plant: a plant from which other plants are started; a source plant for explants.

stock solution: a concentrated solution of media chemicals from which portions are used to make media.

stomate (pl., stomata): a small opening in the epidermis of leaves and stems through which gases and water may pass.

surfactant: a wetting agent that reduces surface tension, thus allowing solutions to better penetrate and clean plants or laboratory surfaces, for example.

suspension culture: *see* **cell suspension culture**.

symbiosis: the living together of two dissimilar organisms, usually to mutual advantage.

systemic: within an organism (system), internal, as a pathogen or systemic weed killer.

tare: to allow for the weight of a container, which is deducted from the total weight of the substance and container being weighed.

tetraploid: 4*n*, double the normal number of chromosomes in vegetative cells.

tissue culture: literally, the culture of individual tissues, but usually used more broadly to indicate micropropagation or *in vitro* propagation.

totipotence: the capability of developing into a whole plant, said of a cell.

transfer: the process of dividing cultures and placing the propagules in containers of fresh sterile medium; also, the piece being so transferred.

transfer chamber or hood: a protected, enclosed area with a sterile atmosphere in which cultures are started, divided, trimmed, and then transferred using sterile technique.

transpiration: the escape of water vapor from a plant, primarily through the leaf stomata.

ultrasonic cleaner: a cleaning device employing ultrasonic vibration, occasionally used for cleaning explants.

undifferentiated: not differentiated, as in cells or tissues that have not yet been modified for their ultimate structure or function.

variegated: plants having multicolored leaves, either due to a virus, a nutritional deficiency, or genetic make-up.

vector: (1) an agent that carries a disease from one organism to another or pollen from one flower to another; (2) in genetic engineering, the plasmid or nucleic acid used to insert genes into receptive cells.

vegetative: somatic, non-sexual.

vitrification: a phenomenon in which tissues develop a glassy, water-logged, swollen appearance.

water of hydration: the amount of water that is chemically attached to a compound. This must be accounted for proportionally if, for example, an anhydrous (lacking the water of hydration) form of a chemical is used instead of the hydrate form.

zygote: a fertilized egg cell; the diploid cell resulting from the union of the male and female sex cells in the process of fertilization.

Bibliography

Abdullah, A. A., M. M. Yeoman, and J. Grace. 1985. *In vitro* adventitious shoot formation from embryonic and cotyledonary tissues of *Pinus brutia* Ten. *Plant Cell, Tissue, and Organ Culture* 5: 35–44.

Almehdi, A. A., and D. E. Parfitt. 1986. *In vitro* propagation of peach: 1. Propagation of 'Lovell' and 'Nemaguard' peach rootstocks. *Fruit Varieties Journal* 40(1): 12–17.

Ammirato, P. V. 1987. Organizational events during embryogenesis. In *Plant Tissue and Cell Culture: The Proceedings of the Sixth International Congress of Plant Tissue and Cell Culture, University of Minnesota 1986*, edited by C. E. Green, D. A. Somers, W. P. Hackett, and D. D. Biesboer. New York: Alan Liss.

Anderson, W. C. 1975. Propagation of rhododendrons by tissue culture: 1. Development of a culture medium for multiplication of shoots. *Proceedings of the International Plant Propagators' Society* 25: 129–135.

———. 1977. Rapid propagation of *Lilium* cv. 'Red Carpet'. *In Vitro* 13: 145.

———. 1978. Rooting of tissue cultured rhododendrons. *Proceedings of the International Plant Propagators' Society* 28: 135–139.

———. 1980. Tissue culture propagation of red raspberries. In *Proceedings of the Conference on Nursery Production of Fruit Plants Through Tissue Culture: Applications and Feasibility*. Washington, DC: U.S. Department of Agriculture SEA Agricultural Research Results.

———. 1984. *Primula* micropropagation multiplies special plants. *Primroses* 42(2): 21–23.

Anderson, W. C., and J. B. Carstens. 1977. Tissue culture propagation of broccoli, *Brassica oleracea* (Italica Group), for use in F_1 hybrid seed production. *Journal of the American Society for Horticultural Science* 102: 69–73.

Anderson, W. C., and G. Haner. 1978. Strawberry tissue culture formula. Mount Vernon, WA: Washington State University, Northwestern Washington Research Unit.

Anderson, W. C., and G. W. Meagher. 1977. Cost of propagating broccoli plants through tissue culture. *HortScience* 12: 543–544.

———. 1978. Cost of propagating plants through tissue culture using lilies as an example. Corvallis, OR: Oregon State University, Ornamentals Short Course.

Anderson, W. C., and K. A. Mielke. 1985. Outline for micropropagating virus-free bulbous iris. Mount Vernon, WA: Washington State University, Northwestern Washington Research Unit.

Anonymous. 1992. *Airborne Particulate Cleanliness Classes in Clean Rooms and Clean Zones*. Washington, DC, FED-STD-209E.

Arditti, J. 1977. Clonal propagation of orchids by means of tissue culture: A manual. In *Orchid Biology: Reviews and Perspectives*, vol. 1, edited by J. Arditti. Ithaca, NY: Cornell University Press.

Arditti, J., E. A. Ball, and D. M. Reisinger. 1977. Culture of flower stalk buds: A method for vegetative propagation of *Phalaenopsis*. *American Orchid Society Bulletin* 46: 236–240.

Ball, E. A. 1978. Cloning *in vitro* of *Sequoia sempervirens*. *Fourth International Congress of Plant Tissue and Cell Culture* (Calgary) 1726: 163.

Ball, V., ed. 1985. *Ball Red Book*. 14th ed. Reston, VA: Prentice Hall.

Barker, P. K., R. A. de Fossard, and R. A. Bourne. 1977. Progress toward clonal propagation of *Eucalyptus* species by tissue culture techniques. *Proceedings of the International Plant Propagators' Society* 27: 546–556.

Barlass, M., and K. G. M. Skene. 1978. *In vitro* propagation of grapevine (*Vitis vinifera* L.) from fragmented shoot apices. *Vitis* 17: 335–340.

Barnett, H. L., and B. B. Hunter. 1972. *Illustrated Genera of Imperfect Fungi.* Minneapolis: Burgess.

Barnett, J. A., R. W. Payne, and D. Yarrow. 1991. *Yeasts: Characteristics and Identification.* 2nd ed. New York: Cambridge University Press.

Beaty, R. M., E. O. Franco, and O. J. Schwarz. 1985. Hormonally induced shoot development on cotyledonary explants of *Pinus oocarpa. In Vitro* 21, Part II: 53A.

Bilkey, P. C., B. McCown, and A. C. Hildebrandt. 1978. Micropropagation of African violet petiole cross-sections. *HortScience* 13: 37–38.

Billings, S. G., C.-K. Chin, and G. Jelenkovic. 1988. Regeneration of blueberry plantlets from leaf segments. *HortScience* 23: 763–765.

Biondi, S., and T. A. Thorpe. 1981. Clonal propagation of forest tree species. *Proceedings of COSTED Symposium on Tissue Culture of Economically Important Plants* (Singapore) 197–204.

Borgman, C. A., and K. W. Mudge. 1986. Factors affecting the establishment and maintenance of 'Titan' red raspberry root organ cultures. *Plant Cell, Tissue, and Organ Culture* 6: 127.

Bottino, P. J. 1981. *Methods in Plant Tissue Culture.* College Park, MD: Kemtec Educational Corp.

Boulay, M. 1979. Multiplication et clonage rapide du *Sequoia sempervirens* par la culture *in vitro. Etudes et Recherches: Micropropagation d'Arbres Forestiers* (Association Forêt-Cellulose) 12: 49–54.

Boxus, P. 1978. Culture de tissus d'espèces fruitières. *Rapport d'Activité Compter Rendus, Gembloux, Belgique* 126–127.

Boxus, P., M. Quoirin, and J. M. Laine. 1977. Large scale propagation of strawberry plants from tissue culture. In *Applied and Fundamental Aspects of Plant, Cell, Tissue, and Organ Culture,* edited by J. Reinert and Y. P. S. Bajaj. New York: Springer-Verlag.

Boyd, C. 1995. Personal communication.

Bridgen, L. H. 1992. A survey of plant tissue culture laboratories in North America. *Proceedings of the International Plant Propagators' Society* 42: 427–430.

Bridgen, M. 1986. Do-it-yourself cloning. *Greenhouse Grower* 1986(May): 43–47.

Briggs, B. 1984. Briggs Nursery, Olympia, WA. Personal communication.

Campbell, R. 1985. *Plant Microbiology.* London: Arnold.

Cassells, A. C. 1991. Problems in tissue culture contaminants. In *Micropropagation: Technology and Application,* edited by P. C. Debergh and R. H. Zimmerman. Dordrecht, Netherlands: Kluwer Academic.

———. 1992. Screening for pathogens and contaminating microorganisms in micropropagation. In *Techniques for the Rapid Detection of Plant Pathogens,* edited by J. M. Duncan and L. Torrance. Boston: Blackwell Scientific.

Chang, H. N., and S. J. Sim. 1994. Increasing secondary metabolite production in plant cell cultures with fungal elicitors. In *Biotechnological Applications of Plant Cultures,* edited by P. D. Shargool and T. T. Ngo. Cleveland, OH: CRC Press.

Chee, R., R. M. Pool, and D. Bucher. 1984. A method for large scale *in vitro* propagation of *Vitis. New York's Food and Life Sciences Bulletin,* No. 109. Geneva, NY: New York State Agricultural Experiment Station.

Chin, C.-K. 1982. Promotion of shoot and root formation in asparagus *in vitro* by ancymidol. *HortScience* 17: 590–591.

Choi, K.-T., M.-W. Kim, and H.-S. Shin. 1982. Root and shoot formation from callus and leaflet cultures of ginseng *(Panax ginseng).* In *Plant Tissue Culture 1982: Proceedings of the Fifth International Congress of Plant Tissue and Cell Culture,* edited by Akio Fujiwara. Tokyo and Lake Yamanaka, Japan: Japanese Association for Plant Tissue Culture.

Christie, C. B. 1985. Propagation of Amaryllids: A brief review. *Proceedings of the International Plant Propagators' Society* 35: 351.

Cooke, L. 1991. Test for toxin-producing molds. *Agricultural Research* 1991(June): 22.

Cooke, R. C. 1977a. Tissue culture propagation of African violets. *HortScience* 12: 549.

———. 1977b. The use of an agar substitute in the initial growth of Boston ferns *in vitro*. *HortScience* 12: 339.

———. 1979. Homogenization as an aid in tissue propagation of *Platycerium* and *Davallia*. *HortScience* 14: 21.

Cousineau, J. C., A. K. Anderson, H. A. Daubeny, and D. J. Donnelly. 1993. Characterization of red raspberries and selection using isoenzyme analysis. *HortScience* 28: 1185–1186.

Curtin, M. E. 1983. Harvesting profitable products from plant tissue culture. *Biotechnology* 1983(October): 649–657.

Dale, P. J., E. Thomas, R. I. S. Brettell, and W. Wernicke. 1981. Embryogenesis from cultured immature inflorescences and nodes of *Lolium multiflorum*. *Plant Cell, Tissue, and Organ Culture* 1: 47–55.

Damiano, C. 1980. Strawberry micropropagation. In *Proceedings of the Conference on Nursery Production of Fruit Plants Through Tissue Culture: Applications and Feasibility*. Washington, DC: U.S. Department of Agriculture SEA Agricultural Research Results.

Davies, D. R. 1980. Rapid propagation of roses *in vitro*. *Scientia Horticulturae* 13: 385–389.

Davis, M. J., R. Baker, and J. J. Hanan. 1977. Clonal multiplication of carnations by micropropagation. *Journal of the American Society for Horticultural Science* 102: 48–53.

Debergh, P. C., and R. H. Zimmerman, eds. 1991. *Micropropagation: Technology and Application*. Dordrecht, Netherlands: Kluwer Academic.

de Fossard, R. A. 1976a. *Tissue Culture for Plant Propagators*. Armidale, Australia: University of New South Wales.

———. 1976b. Vegetative propagation of *Eucalyptus ficifolia* F. Muell. by nodal culture *in vitro*. *Proceedings of the International Plant Propagators' Society* 26: 373–378.

———. 1977. Progress toward clonal propagation of *Eucalyptus* species by tissue culture techniques. *Proceedings of the International Plant Propagators' Society* 27: 546.

———. 1993. Cost of production in commercial micropropagation and research strategies. *The International Plant Propagators' Society: Combined Proceedings* 43: 131–140.

Dimock, A. W. 1943. A method of establishing *Verticillium*-free clones of perennial plants. *Phytopathology* 33: 3.

Dixon, R. A., ed. 1985. *Plant Cell Culture: A Practical Approach*. Oxford: Oxford University Press, IRL Press.

Dixon, R. A., R. B. Boone, A. Garden, L. S. Daley, and T. L. Righetti. 1990. Analysis of electrophoresis isozyme patterns using a microcomputer-based imaging system. *HortScience* 25: 961–964.

Donnelly, D. J., and W. E. Vidaver. 1988. *Glossary of Plant Tissue Culture*. Portland, OR: Timber Press.

Donnelly, D. J., W. E. Vidaver, and K. Y. Lee. 1985. The anatomy of tissue cultured red raspberry prior to and after transfer to soil. *Plant Cell, Tissue, and Organ Culture* 4: 43–50.

Driver, J. A. 1985. Direct field rooting and acclimatization of tissue culture cuttings. *In Vitro* 21: 57A–80.

Driver, J. A., and G. Suttle. 1985. Nursery handling of propagules. In *Cell and Tissue Culture in Forestry*, edited by J. M. Bonga and D. J. Durzsan. Dordrecht, Netherlands: Martinus-Nijhoff.

Dunstan, D. I. 1981. Transplantation and post-transplantation of micropropagated tree-fruit rootstocks. *Proceedings of the International Plant Propagators' Society* 31: 39–44.

Durbin, R. C., ed. 1979. *Nicotiana* procedures for experimental use. Technical Bulletin 1586. Washington, DC: U.S. Department of Agriculture.

Earle, E. D., and R. W. Langhans. 1975. Carnation propagation from shoot tip culture in liquid medium. *HortScience* 10: 608–610.

Feliciano, A. J., and M. de Assis. 1983. *In vitro* rooting of shoots from embryo-cultured peach seedlings. *HortScience* 18: 705–706.

Fitter, M., and A. D. Krikorian. 1985. Mature phenotype in *Hemerocallis* plantlets fortuitously generated *in vitro*. *Journal of Plant Physiology* 121: 97–101.

Fujita, Y. 1985. Efficient production of shikonin derivatives by cell suspension culture of *Lithospermum erythrorhizon*. *In Vitro* 21: 60A.

Fujita, Y., Y. Hara, C. Suga, and T. Morimoto. 1981. Production of shikonin derivatives by cell suspension cultures of *Lithospermum erythrorhizon*. *Plant Cell Reports* 1: 61–63.

Furuya, T., and K. Ushiyama. 1994. Ginseng production in cultures of *Panax ginseng* cells. In *Biotechnological Applications of Plant Cultures*, edited by P. D. Shargool and T. T. Ngo. Cleveland, OH: CRC Press.

Gamborg, O. L. 1986. Cells, protoplasts, and plant regeneration in culture. In *Manual of Industrial Microbiology and Biotechnology*, edited by A. L. Demain and N. A. Solomon. Cambridge, MA: Massachusetts Institute of Technology, Department of Nutrition and Food Science.

Gamborg, O. L., T. Murashige, T. A. Thorpe, and I. K. Vasil. 1976. Plant tissue culture media. *In Vitro* 12: 473–478.

Gasser, C., and R. T. Fraley. 1992. Transgenic crops. *Scientific American* 266(6): 62–69.

Gautheret, R. J. 1942. *Manuel Technique de Culture des Tissus Vegetaux*. Paris: Masson.

———. 1982. Plant tissue culture: The history. In *Plant Tissue Culture 1982: Proceedings of the Fifth International Congress of Plant Tissue and Cell Culture*, edited by Akio Fujiwara. Tokyo and Lake Yamanaka, Japan: Japanese Association for Plant Tissue Culture.

Gebhardt, K. 1985. Development of a sterile cultivation system for rooting of shoot tip cultures (red raspberries) in duraplast foam. *Plant Science* 39: 141–148.

Gelman Sciences. 1993. *Laboratory Filtration*. Ann Arbor, MI: Gelman Sciences.

George, E. F., and P. D. Sherrington. 1984. *Plant Propagation by Tissue Culture*. Eversley, England: Exegetics Ltd.

Glick, B. R., and J. E. Thompson. 1993. *Methods in Plant Molecular Biology and Biotechnology*. Boca Raton, FL: CRC Press.

Goldfarb, B., and G. E. McGill. 1992. *In vitro* culture and micrografting of white pine meristems. *Proceedings of the International Plant Propagators' Society* 42: 412–414.

Goodwin, P. B., and T. Adisarwanto. 1980. Propagation of potato by shoot tip culture in petri dishes. *Potato Research* 23: 445–448.

Gould, J. H., and T. Murashige. 1985. Morphogenic substances released by plant tissue cultures: 1. Identification of berberine in *Nandina* culture medium, morphogenesis, and factors influencing accumulation. *Plant Cell, Tissue, and Organ Culture* 4: 29–42.

Goyal, Y., R. L. Bingham, and P. Felker. 1985. Propagation of the tropical tree, *Leucaena leucocephala* K67, by *in vitro* bud culture. *Plant Cell, Tissue, and Organ Culture* 4: 3–10.

Greenberg, B. M., and B. R. Glick. 1993. The use of recombinant DNA technology to produce genetically modified plants. In *Methods in Plant Molecular Biology and Biotechnology*, edited by B. R. Glick and J. E. Thompson. Boca Raton, FL: CRC Press.

Griesbsach, R. J. 1989. Selection of a dwarf *Hemerocallis* through tissue culture. *HortScience* 24: 1027–1028.

Griffis, J. L., Jr., G. R. Hennen, and R. P. Oglesby. 1980. Establishing tissue cultured plants in soil. *Proceedings of the International Plant Propagators' Society* 33: 618–622.

Griffiths, M. 1994. *Index of Garden Plants*. Portland, OR: Timber Press.

Gugerli, P. 1992. Commercialization of serological tests for plant viruses. In *Techniques for the Rapid Detection of Plant Pathogens*, edited by J. M. Duncan and L. Torrance. Boston: Blackwell Scientific.

Hammerschlag, F. 1982. Factors affecting establishment and growth of peach shoots *in vitro*. *HortScience* 17: 85–86.

Handley, L. W., and O. L. Chambliss. 1979. *In vitro* propagation of *Cucumis sativus*. *HortScience* 14: 22.

Harris, R. E. 1984. Rapid propagation of saskatoon plants *in vitro*. Agricultural Research Council of Alberta, Canada, Farming for the Future Project 83-0054.

———. 1985. Rooting of *in vitro* shoots of saskatoon *(Amelanchier alnifolia)*. Agricultural Research Council of Alberta, Canada, Farming for the Future Project 82-0017.

Harris, R. E., and E. B. B. Mason. 1983. Two machines for *in vitro* propagation of plants in liquid media. *Canadian Journal of Plant Science* 63: 311–316.

Harris, R. E., and J. H. Stevenson. 1982. *In vitro* propagation of *Vitis*. *Vitis* 21: 22–32.

Hartmann, H. T., W. J. Flocker, and A. M. Kofranek. 1981. *Plant Science*. Englewood Cliffs, NJ: Prentice Hall.

Hartmann, H. T., and D. K. Kester. 1983. *Plant Propagation: Principles and Practices*. 4th ed. Englewood Cliffs, NJ: Prentice Hall.

Hartney, V. J. 1982. Tissue culture of *Eucalyptus*. *Proceedings of the International Plant Propagators' Society* 32: 98–109.

Hartney, V. J., and E. D. Eabay. 1984. From tissue culture to forest trees. *Proceedings of the International Plant Propagators' Society* 34: 93–99.

Hasagawa, P. M. 1979. *In vitro* propagation of rose. *HortScience* 14: 610–612.

———. 1980. Factors affecting shoot and root initiation from cultured rose shoot tips. *Journal of the American Society for Horticultural Science* 105: 216–220.

Hennen, G. R., and T. J. Sheehan. 1978. *In vitro* propagation of *Platycerium stemaria* (Beauvois) Deav. *HortScience* 13: 245.

Heuser, C. W., and D. A. Apps. 1976. *In vitro* plantlet formation from flower petal explants of *Hemerocallis* cv. 'Chipper Cherry'. *Canadian Journal of Botany* 54: 616–618.

Heuser, C. W., and J. Harker. 1976. Tissue culture propagation of daylilies. *Proceedings of the International Plant Propagators' Society* 26: 269–272.

Hildebrandt, A. C., A. J. Riker, and B. M. Duggar. 1946. The influence of the composition of the medium on growth *in vitro* of excised tobacco and sunflower tissue cultures. *American Journal of Botany* 33: 591–597.

Hill, S. A. 1984. *Methods in Plant Virology*. Boston: Blackwell Scientific.

Holt, J. G., ed. 1984. *Bergey's Manual of Systematic Bacteriology*. Vols. 1 and 2. Baltimore: Williams and Wilkins.

Horsch, R. B., J. King, and G. E. Jones. 1980. Measurement of cultured plant cell growth on filter paper discs. *Canadian Journal of Botany* 58: 2402–2406.

Hussey, G. 1975. Totipotency in tissue culture explants and callus of some members of the Liliaceae, Iridaceae, and Amaryllidaceae. *Journal of Experimental Botany* 216: 253–262.

———. 1976. Propagation of Dutch iris by tissue culture. *Scientia Horticulturae* 4: 163–165.

Hutchinson, J. F. 1984. Micropropagation of 'Northern Spy' apple rootstocks. *Proceedings of the International Plant Propagators' Society* 34: 38–48.

Jenkins, J. E. E. 1959. Methods for the detection of vascular wilt pathogens in carnation cuttings. *Plant Pathology* 8: 23–27.

Johnson, J. L., and E. R. Emino. 1979. *In vitro* propagation of *Mammillaria elongata*. *HortScience* 14: 605–606.

Jones, J. B. 1986. Determining markets and market potential. In *Tissue Culture as a Plant Pro-*

duction System for Horticultural Crops, edited by R. H. Zimmerman, R. J. Griesbsach, F. Hammerschlag, and R. H. Lawson. Dordrecht, Netherlands: Martinus-Nijhoff.

Jones, J. B., and C. J. Sluis. 1991. Marketing of micropropagated plants. In *Micropropagation: Technology and Application*, edited by P. C. Debergh and R. H. Zimmerman. Dordrecht, Netherlands: Kluwer Academic.

Kaniewski, W. K., and P. E. Thomas. 1988. A two step ELISA for rapid, reliable detection of potato viruses. *American Potato Journal* 65: 561–571.

Kao, K. N., and M. R. Michayluk. 1974. A method for high frequency intergeneric fusion of plant protoplasts. *Planta* 115: 355–367.

Kasperbauer, M. J., R. C. Buckner, and L. P. Bush. 1979. Tissue culture of annual ryegrass × tall fescue F₁ hybrids; callus establishment and plant regeneration. *Crop Science* 19: 457–460.

Kevers, C., and T. Gaspar. 1985. Vitrification of carnation *in vitro*: Changes in ethylene production, ACC level and capacity to convert ACC to ethylene. *Plant Cell, Tissue, and Organ Culture* 4: 215–223.

Kleyn, J., M. Bicknel, and M. Gilstrap. 1995. *Microbiology Experiments: A Health Science Perspective*. Boston: William C. Brown.

Kleyn, J., W. Johnson, and T. Wetzler. 1981. Microbiological aerosols and Anctinomycetes in etiological considerations of mushroom workers' lungs. *Applied and Environmental Microbiology* 41: 1454–1460.

Knauss, J. F. 1976. A tissue culture method for producing *Dieffenbachia picta* cv. *Proceedings of the Florida State Horticultural Society* 89: 293–296.

Knop, W. 1865. Quantitative Untersuchungen uber die Ernahrungsprocesse der Pflanzen. Landwirtsh. Vers. Stn. 70–140.

Knudson, L. 1946. A new nutrient solution for the germination of orchid seeds. *American Orchid Society Bulletin* 15: 214–217.

Krul, W. R., and J. Myerson. 1980. *In vitro* propagation of grape. In *Proceedings of the Conference on Nursery Production of Fruit Plants Through Tissue Culture: Applications and Feasibility*. Washington, DC: U.S. Department of Agriculture SEA Agricultural Research Results.

Kunachak, A., C.-K. Chin, T. Le, and T. Gianfagna. 1987. Promotion of asparagus shoot and root growth by growth retardants. *Plant Cell, Tissue, and Organ Culture* 11: 97–110.

Kunisaki, J. T. 1980. *In vitro* propagation of *Anthurium andraeanum* Lind. *HortScience* 15: 508–509.

Kyte, L., and B. Briggs. 1979. A simplified entry into tissue culture production of rhododendrons. *Proceedings of the International Plant Propagators' Society* 29: 90–95.

Kyte, L., and R. M. Kyte. 1981. Small fruit culture after the test tube. *Proceedings of the International Plant Propagators' Society* 31: 45–47.

Lane, D. W. 1982. *Apple Trees in Test Tubes*. Summerland, BC: Agriculture Canada Research Station.

Lazarte, J. E., M. S. Galser, and O. R. Brown. 1982. *In vitro* propagation of *Epiphyllum chrysocardium*. *HortScience* 17: 84.

Levy, L. W. 1981. A large-scale application of tissue culture: The mass propagation of *Pyrethrum* clones in Ecuador. *Environmental and Experimental Botany* 21: 389–395.

Linsmaier, E. M., and F. Skoog. 1965. Organic growth factor requirements of tobacco tissue cultures. *Physiologia Plantarum* 18: 100–128.

Litz, R. E., and R. A. Conover. 1977. Tissue culture propagation of some foliage plants. *Proceedings of the Florida State Horticultural Society* 90: 301.

Lloyd, G., and B. McCown. 1980. Commercially feasible micropropagation of mountain laurel, *Kalmia latifolia*, by use of shoot-tip culture. *Proceedings of the International Plant Propagators' Society* 30: 421–427.

Loescher, W. H., and C. Albrecht. 1979. Development *in vitro* of *Nephrolepis exaltata* cv. 'Bostoniensis' runner tissues. *Physiologia Plantarum* 47: 250–254.

Maene, L. J., and P. C. Debergh. 1983. Rooting of tissue cultured plants under *in vivo* conditions. *Acta Horticulturae* 212: 335–348.

Makino, R. K., and P. J. Makino. 1977. Propagation of *Syngonium podophyllum* cultivars through tissue culture. *In Vitro* 13: 357.

Matsuyama, J. 1978. Tissue culture propagation of *Nandina. Tissue Culture Association, 29th Annual Meeting* 1978: 357.

Matthews, R. E. F. 1981. *Plant Virology*. New York: Academic Press.

Mauseth, J. 1979. A new method for the propagation of cacti. *Cactus and Succulent Journal* 57: 186–187.

McCown, B., and R. Amos. 1979. Initial trials with commercial micropropagation of birch selections. *Proceedings of the International Plant Propagators' Society* 29: 387–393.

McCullouch, S. M., and B. Briggs. 1982. Preparation of plants for micropropagation. *Proceedings of the International Plant Propagators' Society* 32: 297–303.

McGranahan, G. H., J. A. Driver, and W. Tulecke. 1987. Tissue culture of *Juglans*. In *Cell and Tissue Culture in Forestry*, edited by J. M. Bonga and D. J. Durzsan. Dordrecht, Netherlands: Martinus-Nijhoff.

McGranahan, G. H., and C. A. Leslie. 1988. *In vitro* propagation of mature Persian walnut cultivars. *HortScience* 23: 220.

McGranahan, G. H., W. Tulecke, S. Arulsekar, and J. J. Hansen. 1966. Intergeneric hybridization in the Juglandaceae: *Pterocarya* sp. × *Juglans regia. Journal of the American Society for Horticultural Science* 3: 627–630.

McRae, E. A., and J. F. McRae. 1979. Eight years of adventure in embryo culturing. In *The Lily Yearbook*. North American Lily Society.

McRae, J. F. 1995. The Lily Garden, Boring, OR. Personal communication.

Mele, E., J. Messeguer, and P. Camprubi. 1982. Effect of ethylene on carnation explants grown in sealed vessels. In *Plant Tissue Culture 1982: Proceedings of the Fifth International Congress of Plant Tissue and Cell Culture*, edited by Akio Fujiwara. Tokyo and Lake Yamanaka, Japan: Japanese Association for Plant Tissue Culture.

Meyer, M. M., Jr. 1976. Propagation of daylilies by tissue culture. *HortScience* 11: 485–487.

———. 1980. *In vitro* propagation of *Hosta sieboldiana. HortScience* 15: 737–738.

———. 1982. *In vitro* propagation of *Rhododendron catawbiense* flower buds. *HortScience* 17: 891–892.

Meyer, M. M., Jr., L. H. Fuchigami, and A. N. Roberts. 1975. Propagation of tall bearded iris by tissue culture. *HortScience* 10: 477–480.

Mielke, K. A. 1984. *In vitro* bulbing of bulbous *Iris* cv. 'Blue Ribbon' (*Iris xiphium × Iris tingitana*). Master's thesis, Washington State University, Pullman, WA.

Mielke, K. A., and W. C. Anderson. 1989. *In vitro* bulblet formation in Dutch iris. *HortScience* 24: 1028–1031.

Mii, M., T. Mori, and N. Iwase. 1974. Organ formation from the excised scales of *Hippeastrum hybridum in vitro. Journal of the American Society for Horticultural Science* 49: 241–244.

Mikkelsen, E. P., and K. C. Sink. 1978. *In vitro* propagation of Rieger Elatior begonias. *HortScience* 13: 242–244.

Miksche, J. P. 1994. *The USDA-ARS Plant Genome Research Program*. Washington, DC: U.S. Department of Agriculture.

Miles, C. 1990. Causes and consequences of vitrification *in vitro* and its impact on micropropagation. *BioScience* 1990(22 February).

Miller, G. A., D. C. Coston, E. G. Denny, and M. E. Romeo. 1982. *In vitro* propagation of 'Nema-guard' peach rootstock. *HortScience* 17: 194.

Monette, P. L. 1983. Influence of size of culture vessel on *in vitro* proliferation of grape in a liquid medium. *Plant Cell, Tissue, and Organ Culture* 2: 327–332.

———. 1986. Micropropagation of kiwi fruit using non-axenic shoot tips. *Plant Cell, Tissue, and Organ Culture* 6: 73–82.

Morel, G. 1960. Producing virus-free cymbidiums. *American Orchid Society Bulletin* 29: 495–497.

———. 1964. Tissue culture: A new means of clonal propagation of orchids. *American Orchid Society Bulletin* 33: 473–478.

———. 1974. Clonal multiplication of orchids. In *The Orchids: Scientific Studies*, edited by C. L. Withner. New York: John Wiley & Sons.

Morel, G. M., and J. F. Muller. 1964. La culture *in vitro* du meristeme apical de la pomme de terre. *Compter Rendus de l'Academie des Sciences (Paris)* 258: 5250–5252.

Morris, P., A. H. Scragg, N. J. Smart, and A. Stafford. 1985. Secondary product formation by cell suspension cultures. In *Plant Cell Culture: A Practical Approach*, edited by R. A. Dixon. Oxford: Oxford University Press, IRL Press.

Mudge, K. W., J. P. Lardner, and K. L. Eckenrode. 1992. Effects of lighting and CO_2 enrichment on acclimatization of micropropagated woody plants. In *Proceedings of the International Plant Propagators' Society* 42: 421–426.

Mullis, K. B. 1990. The unusual origin of the polymerase chain reaction. *Scientific American* 262(4): 56–65.

Murashige, T. 1974. Plant propagation through tissue cultures. *Annual Review of Plant Physiology* 25: 135–166.

Murashige, T., M. Serpa, and J. B. Jones. 1974. Clonal multiplication of *Gerbera* through tissue culture. *HortScience* 9: 175–180.

Murashige, T., M. N. Shaba, P. M. Hasagawa, F. H. Taktori, and J. B. Jones. 1972. Propagation of asparagus through shoot apex culture: 1. Nutrient medium for formation of plantlets. *Journal of the American Society for Horticultural Science* 97: 158–161.

Murashige, T., and F. Skoog. 1962. A revised medium for rapid growth and bioassays with tobacco tissue cultures. *Physiologia Plantarum* 15: 473–497.

Nag, K. K., and H. E. Street. 1973. Carrot embryogenesis from frozen cultured cells. *Nature* 245: 270–272.

Nester, E. W., C. E. Roberts, and M. T. Nester. 1995. *Microbiology: A Human Perspective.* Dubuque, IA: William C. Brown.

Nickerson, N. L. 1978. *In vitro* shoot formation in lowbush blueberry seedling explants. *HortScience* 13: 698.

Nitsch, J. P., and C. Nitsch. 1969. Haploid plants from pollen grains. *Science* 163: 85–87.

Oglesby, R. P. 1994. Oglesby Nursery, Hollywood, FL. Personal communication.

Oki, L. R. 1981. The modification of research procedures for commercial propagation of Boston ferns. *Environmental and Experimental Botany* 21: 397–400.

Oliphant, J. 1990. The use of paclobutrazol in the rooting medium of micropropagated plants. *Proceedings of the International Plant Propagators' Society* 40: 358–360.

Oswald, T. H., A. E. Smith, and D. V. Phillips. 1977. Callus and plantlet regeneration from cell cultures of ladino clover (*Trifolium repens* 'Royal Ladino') and soybean (*Glycine max*). *Physiologia Plantarum* 39: 129–134.

Page, Y. M., and J. van Staden. 1984. *In vitro* propagation of *Hypoxis rooperi*. *Plant Cell, Tissue, and Organ Culture* 3: 359–362.

Papachatzi, M., P. A. Hammer, and P. M. Hasagawa. 1980a. *In vitro* propagation of *Hosta plantaginea*. *HortScience* 15: 506–507.

————. 1980b. Tissue culture propagation of *Hosta decorata* 'Thomas Hogg'. *HortScience* 15: 436.

Parfitt, D. E., and A. A. Almehdi. 1986. *In vitro* propagation of peach: II. A medium for *in vitro* multiplication of 56 cultivars. *Fruit Varieties Journal* 40(2): 46–47.

Pasqualetto, P. L, R. H. Zimmerman, and L. Fordham. 1986. Gelling agent and growth regulator effects on shoot vitrification of 'Gala' apple *in vitro*. *Journal of the American Society for Horticultural Science* 111: 976–980.

Peterson, M. A. 1979. Some aspects of nursery production in Queensland. *Proceedings of the International Plant Propagators' Society* 29: 109–110.

Phan, C. T., and P. Hagadus. 1986. Possible metabolic basis for the developmental anomaly observed in *in vitro* culture, called "vitreous plants". *Plant Cell, Tissue, and Organ Culture* 6: 83–94.

Pierik, R. L. M. 1987. *In Vitro Culture of Higher Plants*. Dordrecht, Netherlands: Martinus-Nijhoff.

Pierik, R. L. M., and T. A. Segers. 1973. *In vitro* culture of midrib explants of *Gerbera*: Adventitious root formation and callus induction. *Zeitschrift fur Pflanzenphysiologie* 69: 204–212.

Pierik, R. L. M., H. H. M. Steegmans, and J. A. J. Vak de Meys. 1974. Plantlet formation in callus tissues of *Anthurium andraeanum* Lind. *Scientia Horticulturae* 2: 193–198.

Preece, J. E., and E. G. Sutter. 1991. Acclimatization of micropropagated plants to the greenhouse and field. In *Micropropagation: Technology and Application*, edited by P. C. Debergh and R. H. Zimmerman. Dordrecht: Kluwer Academic.

Preil, W. 1991. Application of bioreactors in plant propagation. In *Micropropagation: Technology and Application*, edited by P. C. Debergh and R. H. Zimmerman. Dordrecht: Kluwer Academic.

Preil, W., and A. Schum. 1985. Experimental scheme for *in vitro* propagation of *Gerbera*. *European Cooperation in the Field of Science and Technology Research*, Cost Project 87. Washington, DC: U.S. Department of Agriculture.

Read, P. E., S. Garton, K. A. Louis, and E. S. Zimmerman. 1982. *In vitro* propagation of species for bioenergy plantations. In *Plant Tissue Culture 1982: Proceedings of the Fifth International Congress of Plant Tissue and Cell Culture*, edited by Akio Fujiwara. Tokyo and Lake Yamanaka, Japan: Japanese Association for Plant Tissue Culture.

Read, P. E., and P. Gavinlertvatana. 1976. Advances in tissue culture: Ray flower and protoplast culture. *Proceedings of the International Plant Propagators' Society* 26: 275.

Rechcigl, M., Jr., ed. 1978. *CRC Handbook*, Series in Nutrition and Food, vol. 3, section G. Cleveland, OH: CRC Press.

Redenbaugh, K. 1990. Application of artificial seed to tropical crops. *HortScience* 25.

Robbins, W. W., T. E. Weier, and C. R. Stocking. 1965. *Botany*. New York: John Wiley & Sons.

Rodriguez, R. 1982. Stimulation of multiple shoot-bud formation in walnut seeds. *HortScience* 17: 592.

Roest, S., and C. S. Bokelmann. 1975. Vegetative propagation of *Chrysanthemum cinerariifolium in vitro*. *Scientia Horticulturae* 1: 120–122.

Sagawa, K., and J. T. Kunisaki. 1982. Clonal propagation of orchids by tissue culture. In *Plant Tissue Culture 1982: Proceedings of the Fifth International Congress of Plant Tissue and Cell Culture*, edited by Akio Fujiwara. Tokyo and Lake Yamanaka, Japan: Japanese Association for Plant Tissue Culture.

Salisbury, F. B., and C. Ross. 1969. *Plant Physiology*. Belmont, CA: Wadsworth.

Schenk, R. U., and A. C. Hildebrandt. 1972. Medium and techniques for induction and growth of monocotyledonous and dicotyledonous plant cell cultures. *Canadian Journal of Botany* 50: 199–204.

Schnabelrauch, L. S., and K. C. Sink. 1979. *In vitro* propagation of *Phlox subulata* and *Phlox paniculata*. *HortScience* 14: 607–608.

Schwann, T. 1839. *Mikroskopische Untersuchungen über die Ubereinstimmung in der Struktur und Wachstum der Tiere und Pflanzen*. Reprinted in 1910 as no. 176 of *Klassiker der exakten Wissenschaften*, by W. Ostwald. Leipzig: Engelmann.

Scully, R. M. 1967. Aspects of meristem culture in the *Cattleya* alliance. *American Orchid Society Bulletin* 36: 103–108.

Seabrook, J. E. A., and B. G. Cumming. 1977. The *in vitro* propagation of amaryllis (*Hippeastrum* spp.) hybrids. *In Vitro* 13: 831–836.

Singh, G., and W. R. Curtis. 1994. Reactor design for plant cell suspension culture. In *Biotechnological Applications of Plant Cultures*, edited by P. D. Shargool and T. T. Ngo. Cleveland, OH: CRC Press.

Skoog, F., and C. O. Miller. 1957. Chemical regulation of growth and organ formation in plant tissues cultured *in vitro*. *Biological Action of Growth Substances Symposium of the Society for Experimental Biology* 11: 118–130.

Smith, E. F., I. Gribaudo, A. V. Roberts, and J. Mottley. 1992. Paclobutrazol and reduced humidity improve resistance to wilting of micropropagated grapevine. *HortScience* 27.

Smith, R. H. 1983. *In vitro* propagation of *Nandina*. *HortScience* 18: 304.

———. 1992. *Plant Tissue Culture: Techniques and Experiments*. San Diego: Academic Press.

Smith, R. H., and A. E. Nightingale. 1979. *In vitro* propagation of *Kalanchoe*. *HortScience* 14: 20.

Smith, W. A. 1981. The aftermath of the test tube in tissue culture. *Proceedings of the International Plant Propagators' Society* 31: 47–49.

Sommer, H. E., C. L. Brown, and P. P. Kormanik. 1975. Differentiation of plantlets in longleaf pine *(Pinus palustris)* tissue cultured *in vitro*. *Botanical Gazette* 135: 196–200.

Sommer, H. E., and L. S. Caldas. 1981. *In vitro* methods applied to forest trees. In *Plant Tissue Culture: Methods and Applications in Agriculture*, edited by T. A. Thorpe. New York: Academic Press.

Song, J. S. 1982. *An Inside Look at Some Economics Concerning Plant Tissue Culture Lab Facilities and Operation*. Chicago, IL: Magenta Corp.

Sonneborn, H. H., M, Dorfner, and E. Bohnke. 1982. Chances for the production of virus-free plants. In *Plant Tissue Culture 1982: Proceedings of the Fifth International Congress of Plant Tissue and Cell Culture*, edited by Akio Fujiwara. Tokyo and Lake Yamanaka, Japan: Japanese Association for Plant Tissue Culture.

Standaert-de Metsenaere, R. E. A. 1991. Economic considerations. In *Micropropagation: Technology and Application*, edited by P. C. Debergh and R. H. Zimmerman. Dordrecht, Netherlands: Kluwer Academic.

Stanley, D. 1995. Tissue-culturing plants on cornstarch. *Agricultural Research* 1995(July): 11.

Steward, F. C., and A. D. Krikorian. 1971. *Plants, Chemicals and Growth*. New York: Academic Press.

Stimart, D. P., and P. D. Ascher. 1981a. Developmental responses of *Lilium longiflorum* bulblets to constant or alternating temperatures *in vitro*. *Journal of the American Society for Horticultural Science* 106: 450–454.

———. 1981b. Foliar emergence from bulblets of *Lilium longiflorum* Thunb. as related to *in vitro* generation temperatures. *Journal of the American Society for Horticultural Science* 106: 446–450.

Stolz, L. P. 1979. Getting started in tissue culture: Equipment and costs. *Proceedings of the International Plant Propagators' Society* 29: 375–381.

Suttle, G. R. L. 1983. Micropropagation of deciduous trees. *Proceedings of the International Plant Propagators' Society* 33: 46–49.

———. 1986. Microplant Nurseries, Gervais, OR. Personal communication.

Takayama, S., and M. Misawa. 1980. Differentiation in *Lilium* bulb scales grown *in vitro*. *Physiologia Plantarum* 48: 121–125.

Takayama, S., M. Misawa, Y. Takashige, and H. Tsumori. 1982. Cultivation of *in vitro* propagated *Lilium* bulbs in soil. *Journal of the American Society for Horticultural Science* 107: 830–834.

Tarr, S. A. J. 1972. *The Principles of Plant Pathology.* London: Macmillan.

Taylor, M. E., and J. F. Knauss. 1978. Tissue culture multiplication and subsequent handling of known pathogen-free *Dieffenbachia maculata* cv. 'Perfection'. *Proceedings of the Florida State Horticultural Society* 91: 233–235.

Thomas, D. des S., and T. Murashige. 1979. Volatile emissions of plant tissue cultures: I. Identification of the major components. *In Vitro* 15: 654–662.

Thompson, D. C., and J. B. Zaerr. 1981. Induction of adventitious buds on cultured shoot tips of Douglas fir (*Pseudotsuga menziesii* [Mirb.] Franco). In *IUFRO Colloque International sur la Culture "in Vitro" des Essence Forestières.* Nangis, France: Association Forêt-Cellulose.

Thorpe, T. A., ed. 1981. *Plant Tissue Culture: Methods and Applications in Agriculture.* New York: Academic Press.

Tombolato, A. F. C., C. Azevedoh, and V. Nagal. 1994. Effects of auxin treatments on *in vivo* propagation of *Hippeastrum hybridum* Hort. by twin scaling. *HortScience* 29: 922.

Tomes, D. T., B. E. Ellis, P. M. Harney, K. J. Kasha, and R. L. Peterson, eds. 1982. *Application of Plant Cell and Tissue Culture to Agriculture and Industry.* Guelph, ON: Plant Cell Culture Centre, University of Guelph.

Torello, W. A. S., and A. G. Symington. 1984. Regeneration from perennial rye grass callus tissue. *HortScience* 10: 56–57.

Tortora, G. J., B. R. Funke, and C. L. Case, eds. 1982. *Microbiology: An Introduction.* Menlo Park, CA: Benjamin-Cummings.

Upham, S. 1983. Plant Resources Institute, Salt Lake City, UT. Personal communication.

U.S. Department of Agriculture Agricultural Research Service. 1994. *The USDA-ARS Plant Genome Research Program.* Athens, GA: University of Georgia.

Vacin, E. F, and F. W. Went. 1949. Some pH changes in nutrient solutions. *Botanical Gazette* 110: 605–613.

van Aartrijk, J., and P. C. G. van der Linde. 1986. *In vitro* propagation of flower-bulb crops. In *Tissue Culture as a Plant Production System for Horticultural Crops*, edited by R. H. Zimmerman, R. J. Griesbsach, F. Hammerschlag, and R. H. Lawson. Dordrecht, Netherlands: Martinus-Nijhoff.

van Overbeek, J., M. E. Conklin, and A. F. Blakeslee. 1941. Factors in coconut milk essential for growth and development of very young *Datura* embryo. *Science* 94: 350–351.

Vasil, I. K., ed. 1984. *Cell Culture and Somatic Cell Genetics in Plants: Laboratory Techniques.* New York: Academic Press.

von Arnold, S., and T. Eriksson. 1981. *In vitro* studies of adventitious shoot formation in *Pinus contorta. Canadian Journal of Botany* 59: 870–872.

Wang, A. 1990. Callus induction and plant regeneration of American ginseng. *HortScience* 25: 5.

Wehner, T., and R. D. Locy. 1981. *In vitro* adventitious shoot and root formation of cultivars and lines of *Cucumis sativus* L. *HortScience* 16: 759–760.

Wetherell, D. F. 1982. *Introduction to in Vitro Propagation.* Wayne, NJ: Avery.

White, P. R. 1943. *A Handbook of Plant Tissue Culture.* Tempe, AZ: Jaques Cattell.

———. 1963. *The Cultivation of Animal and Plant Cells.* 2nd ed. New York: Ronald.

Wilkins, M. B. 1984. *Advanced Plant Physiology.* London: Pitman.

Wimber, D. E. 1963. Clonal multiplication of cymbidiums through tissue culture of the shoot meristem. *American Orchid Society Bulletin* 32: 105–107.

Windholz, M., et al., ed. 1989. *The Merck Index.* 11th ed. Rahway, NJ: Merck & Company.

Wong, S. 1981. Direct rooting of tissue cultured rhododendrons into an artificial soil mix. *Proceedings of the International Plant Propagators' Society* 31: 36–37.

Wood, M. 1989. Blast those genes. *Agricultural Research* 1989(June).

Yang, H. J., and W. J. Clore. 1973. Rapid vegetative propagation of asparagus through lateral bud culture. *HortScience* 8: 141–143.

———. 1974. Development of complete plantlets from vigorous shoots of stock plants of asparagus *in vitro*. *HortScience* 9: 138–140.

Yanny, D. 1988. A study to determine the more economically viable alternatives for greenhouse operations between micropropagation and lateral shoot production of plants. Bachelor's thesis, Cardinal Stritch College, Wisconsin.

Young, P. M., A. Hutchins, and M. L. Canfield. 1984. Use of antibiotics to control bacteria in shoot cultures of woody plants. *Plant Science Letters* 34: 203–209.

Zenk, M. H., H. El Shagi, H. Arens, J. Stockigt, E. W. Weiler, and B. Deus. 1977. Formation of the indole alkaloids serpentine and ajmalicine in cell suspension cultures of *Catharanthus roseus*. In *Plant Tissue Culture and its Bio-Technological Application*, edited by W. Barz, E. Rienhard, and M. H. Zenk. Berlin: Springer-Verlag.

Zillis, M., and D. Zwagerman. 1979. Clonal propagation of hostas by scape section culture *in vitro*. *HortScience* 14: 80.

Zillis, M., D. Zwagerman, D. Lamberts, and L. Kurtz. 1979. Commercial propagation of herbaceous perennials by tissue culture. *Proceedings of the International Plant Propagators' Society* 29: 404–413.

Zimmerman, R. H. 1978. Tissue culture of fruit trees and other fruit plants. *Proceedings of the International Plant Propagators' Society* 28: 539–546.

———. 1986. Propagation of fruit, nut, and vegetable crops: An overview. In *Tissue Culture as a Plant Production System for Horticultural Crops*, edited by R. H. Zimmerman, R. J. Griesbsach, F. Hammerschlag, and R. H. Lawson. Dordrecht, Netherlands: Martinus-Nijhoff.

———. 1991. Micropropagation of temperate zone fruit and nut crops. In *Micropropagation: Technology and Application*, edited by P. C. Debergh and R. H. Zimmerman. Dordrecht, Netherlands: Kluwer Academic.

Zimmerman, R. H., and O. C. Broome. 1980a. Apple cultivar micropropagation. In *Proceedings of the Conference on Nursery Production of Fruit Plants Through Tissue Culture: Applications and Feasibility*. Washington, DC: U.S. Department of Agriculture SEA Agricultural Research Results.

———. 1980b. Blueberry micropropagation. In *Proceedings of the Conference on Nursery Production of Fruit Plants Through Tissue Culture: Applications and Feasibility*. Washington, DC: U.S. Department of Agriculture SEA Agricultural Research Results.

———. 1980c. Micropropagation of thornless blackberry. In *Proceedings of the Conference on Nursery Production of Fruit Plants Through Tissue Culture: Applications and Feasibility*. Washington, DC: U.S. Department of Agriculture SEA Agricultural Research Results.

Ziv, M. 1986. *In vitro* hardening and acclimation of tissue culture plants. In *Plant Tissue Culture and its Agricultural Applications*, edited by L. A. Withers and P. G. Alderson. London: Butterworth.

Index

Actinidia 197
aeration 83, 104
African violet 192
agar 45, 71, 109
air quality 37, 42, 52, 90, 114, 123–126
alcohol 48, 88, 90
algae 157
amaryllis 171
Amelanchier 198
anther culture 32, 156
Anthurium 173
antibiotics 71
apple 199
Arabidopsis 153
arrowhead vine 174
Asparagus 176
atomic weights 62
auxins 31, 69, 78, 103, 109

bacteria 29, 101, 113, 118, 120, 121, 126, 128
Begonia 76, 184
bioreactors 160
blackberry 202
bleach 50, 87, 89, 90
bleeding. *See* phenolics
blueberry 192
Brassica 189
bridges 82
broccoli 189
Bryophyllum. See *Kalanchoe*
bulbs 23

cacti 186
callus 16, 22, 25, 30, 32, 96, 153, 155
carnation 187
carrot 30, 31, 33, 153, 155
Catharanthus 183
Cattleya 179
cell culture 155
charcoal 72, 103
cherry 200
chromosomes 25. *See also* haploids; meiosis; mitosis
Chrysanthemum 188
coconut milk 31, 72
cold storage 51, 105. *See also* cryopreservation
computers 138, 143
conifers 157, 170
costs 101, 138, 140

contamination 141
labor 17, 140, 141
space 17
cryopreservation 158
cucumber 190
Cucumis 190
customs. *See* shipping
cuttings 16, 17, 21, 31. *See also* microcuttings
Cymbidium 181
cytokinins 31, 69, 79, 103, 109

Daucus 30, 31, 33, 153, 155
daylily 177
Dianthus 187
Dieffenbachia 173
differentiation 25
disinfectants 37, 48, 49. *See also* alcohol; bleach; hydrogen peroxide
DNA (deoxyribonucleic acid) 147–153
dumb cane 173

electrophoresis 153
ELISA. *See* indexing
embryo culture 30, 31, 32, 97
embryogenesis, somatic, 32, 155, 157, 194
embryo rescue 30, 31
endospores 29, 113
enzymes 147, 148, 149
Epiphyllum 186
Eucalyptus 195

ferns 169
fescue 174
Festuca 174
filters 37
furnace 37, 43
HEPA 37, 52
membrane 124
fingerprinting. *See* electrophoresis
flamingo flower 173
forests 157
Fragaria 94, 97, 198
Freesia 175
fungi 118, 128, 161
fungicides 111

Gerbera 188
gibberellins 70
ginseng 183

grafting 23. *See also* micrografting
grape 203
growing room 36, 37, 101, 110
growth regulators. *See* hormones
guttation 108

haploids 32. *See also* meiosis.
haploid culture 32, 156
heat treatment 84
heavenly bamboo 185
Hemerocallis 177
HEPA filters. *See* filters, HEPA
Hippeastrum 171
hood. *See* transfer chamber
hormones 68, 103. *See also* auxins; cytokinins; gibberellins
Hosta 178
hydrogen peroxide 88
Hypoxis 172

indexing 102, 129, 132. *See also* viruses
Iris 175
irradiance. *See* lighting

Jiffy 7's 112
Juglans 193
juvenility 85, 157

Kalanchoe 189
Kalmia 190
kiwi fruit 197

labeling 51, 91, 94
laboratories 13, 16, 20, 35, 137, 139
laminar air flow hood. *See* transfer chamber
laurel 190
layering 22
air layering 23
lead tree 195
leaf 24, 25, 84, 86, 156
as explant 86
leaf mesophyll 156
maceration 84, 96
Leucaena 195
lighting 37, 38, 39, 101
Lilium 178
lily 178
plantain lily 178
limiting factors, theory of, 30
Lithospermum 185
Lolium 174

Malus 199
Mammillaria 186
media 60, 76, 98, 102. *See also* agar.
    Anderson's 179, 191, 197
    dispensing 47, 81
    DKW (Driver/Kuniyuki/ Walnut) 193
    Gamborg's B5 183
    indexing 129, 132
    liquid 50, 82
    MS (Murashige and Skoog) 32, 77, 168
    Nitsch's 196
    premixed 60, 76, 140
    undefined 71
    Vacin and Went's 181
    White's 186, 194
    WPM (woody plant medium) 60, 190
media preparation room 36, 39
meiosis 25, 149
meristem 25, 129
    apical meristem 25
    meristem culture 94, 98
metabolites, secondary. *See* secondary products
metric system 61
microcuttings 22, 108
micrografting 23, 97
microscope 27, 29, 51, 118, 133
Miltonia 181
mitosis 25, 149
mold. *See* fungi
molecular weight 62
mutation 16, 25, 26, 34, 154

Nandina 185
Nephrolepis 169
Nicotiana 31, 32, 33, 133, 153

Odontoglossum 181
Oncidium 181
orchids 13, 31, 72, 97, 179. See also *Cattleya*; *Cymbidium*; *Miltonia*; *Odontoglossum*; *Oncidium*; *Phalaenopsis*
organogenesis 155, 158

Panax 183
periwinkle 183
pH 44, 63

Phalaenopsis 182
phenolics 72, 102
phloroglucinol 70, 105
Phlox 196
photoperiod. *See* lighting
pine 97, 157, 171
Pinus 97, 157, 171
plantain lily 178
plasmid 149, 151
plum 200
potato 203
premixes. *See* media, premixed
Primula 197
production 16, 17, 18, 137, 138, 140. *See also* costs.
    planning 142
    time 144
protoplasts 33, 152, 156
Prunus 200
pyrethrum 188

quality control 18, 114, 115

rafts. *See* bridges
raspberry 202
record keeping 143
red-flowering gum 195
redwood 170
refrigeration. *See* cold storage
resin bed 43
rhizomes 23
Rhododendron 97, 98, 99, 191
RNA (ribonucleic acid) 147
roots 23, 24
    aerial 25
    culture of 25, 69
    initiation (rooting) 69
    stock 22, 23, 112
    tips 87
Rosa 201
rose 201
Rubus 202
runners 95, 198
ryegrass 174

Saintpaulia 192
secondary products 34, 159, 183, 185
seedlings 15, 108
seeds 15, 16, 17, 25, 97
Sequoia 170
serviceberry 198

shikonin 159, 160, 185, 186
shipping 145
    customs 35, 139
soil, soil mixes, 108, 112
Solanum 203
somatic embryogenesis. *See* embryogenesis, somatic
spontaneous generation, theory of, 28
staghorn fern 169
stains, microbial, 121
star grass 172
sterile technique 90
stock cultures 105
stock plants 16, 17
stock solutions 76
stolons 23
storage. *See* cold storage; stock cultures
strawberry 94, 97, 198
sucrose (sugar) 67
suspension culture. *See* media, liquid
Syngonium 174

temperature 39, 42, 85, 110
tobacco 31, 32, 33, 133, 153
totipotence, theory of, 29
transfer chamber (hood) 29, 36, 52, 90
transferring cultures 35, 90, 93, 97, 98
transfer room 36, 90
transpiration. *See* water, loss
Transvaal daisy 188
trays, planting, 55, 109, 110
tubers 23

Vaccinium 192
vegetative reproduction 15, 21, 25
viruses 94, 113, 132, 133, 147. *See also* indexing
Vitis 97, 203
vitrification 104

walnut 193
water 42, 110
    loss 108
    quality 42
wounding 22, 25

yeast 30, 72, 113, 118, 120, 121, 129